T0064197

PUZZLES *of* HISTORY

PUZZLES *of* HISTORY

Truths and Untruths Part-I

SUDIP NARAYAN GHOSH

PARTRIDGE
A Penguin Company

Copyright © 2013 by Sudip Narayan Ghosh.

ISBN:	Hardcover	978-1-4828-1153-7
	Softcover	978-1-4828-1154-4
	Ebook	978-1-4828-1152-0

All rights reserved. No part of this book may be used or reproduced by any means, graphic, electronic, or mechanical, including photocopying, recording, taping or by any information storage retrieval system without the written permission of the publisher except in the case of brief quotations embodied in critical articles and reviews.

Because of the dynamic nature of the Internet, any web addresses or links contained in this book may have changed since publication and may no longer be valid. The views expressed in this work are solely those of the author and do not necessarily reflect the views of the publisher, and the publisher hereby disclaims any responsibility for them.

Published by Partridge Publishing India

Partridge books may be ordered through booksellers or by contacting:

Partridge India
Penguin Books India Pvt.Ltd
11, Community Centre, Panchsheel Park, New Delhi 110017
India
www.partridgepublishing.com
Phone: 000.800.10062.62

DEDICATED

To
My Mother
Late Avarani Ghosh

PREFACE

The story of mankind since it developed from its primitive stage of barbarism and savagery to one of civilization has been traced as far back as five thousand years before Christ. The formal documentation of human history started with Herodotus. The method and techniques applied for study of history have since changed with the advancement of science and technology. Given the limitations for chronicling of history it is possible to encompass only a little of myriad dimensions of this grand and internecinely varying life. Despite whatever precious little could be achieved certain errors and discrepancies of substantial nature have crept into our history and collective psyche both at the national and international level. There are an amazingly large number of issues and thorny problems that have no explicit answer. It is wise to accept those conflicting ideas.

I am not that versatile to properly reproduce and verbalize the ripples rising in my mind. With modest exposure in this subject I have relied mainly on my common sense and personal experiences marshaled here and there. In doing so if any lapses occur and should anyone feel otherwise I shall gladly invite his/her advice and suggestions to set myself corrected. Queries that may crop up in inquisitive mind shall be attempted to be clarified in the subsequent Parts.

SUDIP NARAYAN GHOSH

CONTENTS

INTRODUCTION

When in the distant and obscure past human race made its appearance on the virginal earth he was bewildered and confused. Nature's fury and ravages wrought in its train left him with no clue as to the ways of escape from this horror and human brain, we are not confirmed about the different chronological stages through which it developed, very inadeptly could handle the situation. As the streak of lightning pierced through rumbling layers of dark cloud, horizon to horizon, with heart rending loudness, deluge swept away human lives with merciless ferocity, cyclonic wind blew apart the thin slices of their home and hearth, volcanoes spewed fire and lavas rolled down churning anything that came in the way innocent men and women looked agape. The most natural reaction was to drive him to the conclusion that some evil agency was at work from behind the screen. Thus the concept of something unknown or superior power germinated in the mind not in the form of benevolent Supreme Being but in the form of an evil-doer.

Against this harmful authority it was necessitated to set an opposite form of superior force that is able to contend with the former in order to ensure peace, security and stability to man. The primitive man without any prejudice and a legacy of hardened ideas almost cornered to the wall formed hazy ideas about the possibilities of an extant and intractable being that can wield enormous strength on the affairs of this happening world. Since he could not fathom the depth and extent of this super-natural, invisible and ominous force and it was equally unacceptable for him to conceive of a situation where he has to suffer ignominious defeat at the hands of nature, due to an unspecified feeling of survival against oddities, that came to be identified as the resultant effect of ego, the principal motive force behind all human activities, he started building an extra-terrestrial world with a benign and superior being at the helm who will finally give him peace and happiness.

Nature's fury apart he was continually hounded by fierce animals, venomous reptiles and more strikingly he was subject to vandalisation by fellow humans, the trace of which are still apparent in the form of war and riots in varying degrees. With the passage of time as a by-product of intelligence sense of belongingness and possessiveness developed; desires and aspirations grew on individual and group levels. But it also dawned on them that fulfillment of these ambitions and contentment of inner-self are unattainable, their capacity or control over the result of their attempts being vigorously limited by nature. It, as it transpired, was almost impossible to foresee or foretell what is in store for them in future or what lie ahead even in the next couple of hours, it became almost a painful compulsion to conjecture and fancy something larger than common life that will take care of all these uncertainties and fearsome antagonists.

Another aspect that concerns us or disturbs our mind is the contradiction within our own self resulting from our relations or interactions with the external world. As we cannot remain alienated and detached from our fellow beings, not because of others' satisfaction but for own ego our behavior towards them vis-à-vis theirs' towards us raises storms within our inner self and torments us constantly. Man, to put it simply, cannot live but without the contact of another man because of the same reason as he comes into clash with them. A man who is deported to a desolate land isolated from all human contact but at the same time is provided with all the necessaries of life will become mentally disoriented ultimately leading to insanity. This is inevitable as he wants to be recognized by another human being eager to quench his thirst to satiate his ego. That his very existence remains unnoticed irks him until he can speak his heart out that is unequivocally reflected in the mirror of another man. He likes to share his emotions and sentiments, toils and turmoil, grief and sorrow like in the same way as he wants to partake his joys and ecstasies, he yearns to enjoy rejoicing with others; the complex and internecinely varying structure of human relations thus grow on a very palpable tone.

Besides the diktat of civilization that enjoins us to live a collective social life, as we have seen in the preceding lines, it is an irrepressible inner urge flowing outwards that drives man to build one to one or one to many relations. In course of this man to man relation, individual or societal, there is continuous exchange of love and hatred, acceptance and rejection, respect and insult. In this constant flow of tides of emotion one man's action , right or wrong , may or may not get the nod of the other . The former, if not, sits on judgment, mistakenly very often though, for the deeds or presumably misdeeds of the other. He passes the verdict, based on his standards or knowledge of the existing moral, legal or religious ordains, unilaterally or by the inductive logic of the mass of evidence influenced by others.

But in hindsight, as he starts recounting he quite invariably stumbles and abruptly explores in the process that what he thought had human or moral sanction was essentially not that human. And finally realizes that the question of humanity is much more complicated than it is usually perceived to be as per existing laws or notions erected over a long period looked through the canons of established religious or social parameters. Here comes the intractable confusion—which one to give precedence—his own judgment rebutting other's or action of others he judged—and embroils him in an endless battle within, lands him in a vortex of right or wrong paradox.

The simmering rage places him in an unanswerable puzzle—'Do I have the right or competence to determine the propriety of other's action?' It leads him to search an escape route from this internal strife unshared by anyone else and the only solution he finds in amazement that he has to repose his job and give his bruised heart some solace by transferring all his queries to a superior authority that has his benign and careful gaze over every temporal and celestial affairs. Whether this solution is sane or insane is altogether a different issue but more importantly it is the only answer to his tortuous feelings.

Apart from the conflict arising from inter-personal relations another conflict confined to lesser area but by no means less disturbing is the contention stemming from the effects on a person due to his own deeds. More so often than not it is presumed, a man commits or omits something and in retrospect he finds that because of his own words or deeds he has placed himself in a very uneasy situation and he has been stripped of the happiness and achievements which he deserved in all likelihood. He reflects that though he is not that foolish the way he has acted and though he was fully capable of discreet judgment with all the analytical quotients available he has acted in a manner that has brought his downfall and degradation irretrievably. His sufferings he cannot expatiate to others and he is left all alone grumbling and stupefied. Once again he falls back upon an imaginary entity at the charge of all good or bad, right or wrong—and concludes that this is the design of that almighty—the God. These internal conflicts in man have been extensively dealt with by Tagore in his poems and dramas with superb literary finesse and dexterity. We will take up this topic in details in subsequent chapters.

We are not going to deal with an antiquated subject as the existence of God or theology; but we will try to arrive at a greater truth—whether what are passed on in the name of historical truth and make-believe truth are acceptable with any amount of reasonableness and how they fog our vision and adversely affect our life and the pursuit of truth itself. Concept of God or its illusive ramifications distorts our realization of history and estimate of mankind in general.

———◆———

CHAPTER 1

Oh God!

The three-lettered word "God" or its counterpart in other tongues is the most popular and widely pronounced one in the world. It has dominated all our thoughts. It has initiated philosophical discourses, influenced our literature, produced volumes of poems and songs and affected our daily life. Notwithstanding anything else it recoils and boils down to a simple inquisition—"Whether God exists or not?" To this there are an amazingly large number of answers without any decisive and all-conclusive solution.

It is inarguably true that an overwhelmingly large number of people irrespective of religious affiliations, cultural tradition or racial denomination steadfastly believe in the existence of god. Some even go to the extent that God only exists and nothing else. They are broadly bracketed as theists-believers in God. It is also undeniable that not an insignificant fraction of people though minority but endowed with high intellectual and moral strength do not have faith in God. They are atheists or non-believers. As opposed to these two extremes there is a third one in between-Agnostics who hold that nothing can be known about the existence of God.

Now it is time to ponder whether there is any possibility of such an eventuality that God, as the normal connotation imports, in any form or hues exists. In this respect all the religious texts or essays on religion that incessantly sing in praise of God maintain a queer silence. Nowhere do we find any trace or proof of the existence of God unquestionably. The believers in God exhibit a peculiar common feature—whenever they are asked 'What is the proof or evidence that god is there?' they fly into rage and hurl invectives about the silliness of such a question. To raise any doubt or make any oblique reference to the impossibility of any mystic occult or unsavory things that are passed in the name of God or its associate religion constitute a crime and punishable as sacrilege. The infidelity is frowned upon and such infidels are persecuted in many ways.

The initial sloka of Upanishad says that *this universe is surrounded by God*. Nothing is clear whence God came or what its form is. We come to know of the various manifestations of the material world through different organs which are governed by our sensory—nervous systems. Though limitations are there in our capacity to conceive of the extremely complex nature of each aspect of this universe, it cannot be an argument in favour of granting us a license to assume in wild fantasy, any absurd and irrational things.

Who has created God? If God created whatever we know or whatever we do not but may exist as perceived to be possible to exist by extension of our experience and knowledge, who created God? This is not a novel query we are making. Given the premise that a vast number of mysteries of this creation are yet to be unraveled, we have to go far to discover the ultimate truth of this infinite universe. We are yet to know which the smallest particle of matter is or how the numbers of forces that regulate the activities of gigantic celestial bodies and that which bind together the fundamental particles within the atom and at the sub-atomic level are interrelated.

Efforts are on to construct theories to converge together all the forces viz., gravitational force, electromagnetic force and nuclear force etc. and form a unified theory that can explain and put at rest all curiosities of mankind about the origin of creation of this universe. We are equally unable to arrive at a conclusive decision whence life came and the intricate mechanism that control the biotic world.

The cumulative experience of man has taught that the best system developed so far to unfold the truth about the biotic and abiotic world is different branches of science. Limitations notwithstanding there are no better alternative. Scientific truth is not accepted, it is the merit of science that scores over other branches of knowledge, until it is established by experiments carried out on assumption of truthfulness of a scientific law or hypothesis.

Avogadro postulated that under the same temperature and pressure equal volumes of all gases contain an equal number of molecules. He did not count the number which is impossible for any human being as the size of molecules is infinitesimally small and number is inordinately large. That proposition being impish scientists deduced formulae on the basis of his predictions which could be put to tests in laboratory. The results of the tests indisputably established the genuineness of his assumptions. Likewise in biological science also the correctness of Gene theory by Creek, Watson and others of the ilk could be proved through various experiments.

If I am asked whether God exists or not I am to say 'no' irrespective of what great thinkers and eminent persons of the world may understand—be it Marx or Einstein, Gandhi or Tolstoy, Thomas Mann or Tagore. There are, other than the renowned personalities as mentioned, innumerable persons though less known but not of mean intellect and negligible faculty who may or may not believe in God or its Avatars or incarnates.

One point must be clarified at the outset. All our problems spring from our inability to speak in straight clean-cut language. This is abundantly clear that not an iota of evidence of the existence of God has been furnished by any religionists or great men of the world. Some blindly express faith in God while others make some bizarre comments which create further confabulations. These points we will take up later on.

Our main concern is not the existence of God to which our answer is affirmatively "no". But our liability to mankind or its ceaseless and irrepressible queries about the wonder that the universe is does not end with a simple statement—'God does not exist'. To a serious thinker the question must recoil—why do millions of people all over the world including the best brains believe in God? Are all of them mad or do they suffer from some kind of mental disorder?

There is, in spite of differences in religious customs and practices, national heritage, historical or social background and cultural features, a unity in the realm of thought of mankind. If one goes through the literatures written in different languages spread over the globe the unique feature and the commonness of mankind that is revealed is this—wherever the issue relates to basic and inmost feelings all are equal. All men or women, the old and the infant think the same way.

The lullaby expresses the same emotion be it in Spanish or Bengali, Japanese or English. These short forms of rhymes or songs are sung by mothers of all the languages in inharmonious voices to pacify their babies to sleep. Surprisingly there are striking similarities in the sentiments and emotions carried by these songs. These songs are never written by any conscious efforts of poets worth the name. Their origin is unknown. Rather they are invariably composed orally by rural women by and large illiterate.

Important it is to note that these songs are handed down from an unknown past, verbally generation to generation. The harmonious chord transmitted through spontaneous outpourings of love and a tensile emotional exposition definitely underscores an eternal truth. A Spanish mother in all probability didn't have connection with her Bengali counterpart so that either of them could copy from the other. The most logical inference that can be drawn is this—there is a nature-dictated coherence in the pattern of thoughts of all human being belittling their geographical boundaries.

*   *   *

It is not possible for me to place irrefutable faith on God as the creator, regulator and the savior of this world, despite that I am not a stark materialist rather I refuse to be a materialist at all. The whole gamut of human entity will be captured in the design of physiology and psychology—it is too difficult a proposition to hold on to. Even it is not quite sure that Physics, the fundamental of all sciences will be able to picturise or visualize the material world in its entirety. In spite of its limitations persuasions of knowledge through scientific method is the best discovery of human brain. Yet with due regard to science we cannot forget that only a portion of this multi-dimensional world is covered by science, the rest is beyond our perception and shrouded in mystery.

Here I have a case to fight with the deadwood materialist who pooh-poohs at the slightest mention of God or anything beyond mundane affairs of life. To them as Charvak said-"So long you live, live in happiness, borrow money and eat cream" (forget the inadequacy of my translation)—is the gist of life. To them all these talk of mystery and mysticism are bogus and need not draw attention of any sensible person with minimal intelligence. Life, as they consider. is nothing more than as we see it with our five organs. We cannot write them off or scorn at them. To them our answer is 'yes' and 'no' both.

Yes—there is no concrete or pliable or plausible proof that there is anything like God or God—related myths.

Yes, by pooling or mustering the entire rationale one cannot accept the existence of a superior power.

No—, in the psychological arena nothing is visible. Can we deny the act of mind, the wide spectrum of emotions, the interplay of light and shadow?

No, we cannot explain our fits of affection and animosity, indolence and humility, cruelty and compassion with organic reality. There is no testimony, as yet of the origin of life. It is of great doubt whether we can break the barriers of fundamental particles or discern the origin of energy. All these guide us to think beyond the realm of this discrete reality and tread in some other path.

> *"Whatever character our theology may ascribe to him (God) in reality he is the infinite ideal of Man towards whom men move in their collective growth"—Religion of Man—Tagore.*

Elsewhere he stated that an animal is an animal since birth but a man has to develop into a man. By origin man's animist instincts are latent which very often express them under specific circumstances. At the same time a different kind of penchant for attaining higher ideals of manhood urges him towards pursuits of various kinds of activities invigorated by those ideals. He does not feel satisfied with his achievements and hence does not stop at that. He almost fanatically endeavors to scale the peak of perfection which is but an enigma. But in his wild and intense fancy to reach that pinnacle of completeness, entirety or faultless wholesomeness he knows no rest, no complacence till that goal is achieved. In his insatiable thirst for that paragon of beatitude which he feels unattainable he builds an imagery of perfectionism. It is this ideal of Man which Tagore termed as 'infinite' towards which mankind marches not as a disparate entity but as a collective whole.

God, according to him is the personification of that infinite (that is why unattainable) ideal (not real in usual objective sense, neither discernible as it pertains to higher echelons of meditative thoughts of Man). An avid reader cannot miss the capital letter used in '**Man**'. Here Man itself is elevated to a higher level akin to God. At no point Tagore, by any stretch of argumentation, dropped a hint for the existence of God. Rather he deftly defined God in most refined term in renunciation of the one claimed by sizeable chunk of popular theists.

In times of unrest and turmoil masses are prone to hatred and cruelty, whereas in times of peace these traits of human nature emerge but stealthily. Einstein wrote about Psalms (a devotional song or hymn especially those included in the Old Testament)—"*a sort of intoxicated joy and amazement at the beauty and grandeur of this world of which man can form just a faint notion. This joy is the* feeling from which true scientific research draws its spiritual sustenance but which also seems to find expression in the song of birds. To tack this feeling to the idea of God seems mere childish absurdity." He further went on—"*The life of the individual only has meaning in so far as it aids in making the life of every living thing nobler and more beautiful. Life is sacred, that is to say it is the supreme value, to which all other values are subordinate.*" Here we have a voice again suggestive of making life of entire human race nobler and that according to Einstein is of supreme value superior to others.

> "This firm belief (a conviction, akin to religious feeling of the rationality or intelligibility of the world lies behind all scientific work of a higher order), <u>a belief bound up with deep feeling, in a superior mind that reveals itself in the world of experience</u>, represents my conception of God. In common parlance, this may be described as pantheistic."

CHAPTER 2

Death and Dying

The worst and inevitable enemy of man is death. Perhaps it is the only killer. Most of philosophical thoughts originate from this sense of death. The first onslaught in realizing the significance of life comes from death. What is the absolute value of life if it meets with abrupt and irreversible end with death? So it became imperative to devise a way out in order to provide solace to the wounded mind. It was proposed that only with the realization of whole indivisible entity, call it Brahma, (Allah or God?) man can transcend the barriers of death. A disparate human individual shall perish without exception. But the entire human race, with its annals without beginning or end is immortal. *'I shall die but we humans shall never be finished forever.'*—this conviction is immortality and herein lies impregnated the value of life. The enterprise which is valueless for an individual is of supreme value in the long term existence of humanity. From this sense of larger or greater 'self' evolved the idea of *Atma* or soul.

The complete annihilation of individual self and the utter and apparent derecognition of all the significance of human life is a defeat to the survivors. In various ways man has therefore tried to say that everything does not end with death. And somehow in order to placate the bereaved heart the theory of rebirth has been forwarded to mean that the erosion in death is only transitory and relative.

Maitrayi asked, 'What use is it of if I do not earn immortality?' *Yajnyavalkya* advised, "You place your own self in the world." That fortitude including simultaneously the material and spiritual world, eternal and temporal, infinite and finite, transcendent and immanent is soul, it is the Brahma. Alternatively it is that knowledge which helps man transcend death and makes him immortal for individual dies not the human race. Man's journey beginning from the earliest of its history towards an unknown future has no end. He is the 'I', or 'Ego', he is the 'Brahma' and in man's emotional term it is the '*Atma* 'or soul. With this knowledge man gets solace that in the backdrop of endless and vast universe he is indestructible, immortal after death.

Man bounded within the tiny cells of time ad space, fragmented by nationhood, culture and customs, suffering from disease, senility and death tormented by want, meanness and ugliness attain a semblance of immortality in that concept, that imagery of Manhood in eternal humanity.

As yet it could not be established that there is any super power or god existing somewhere in heaven or hell (which in itself is non-existent or unreal) who oversees all the affairs of this planet and extra-planetary universe. Neither is there anything in reality called 'Atma' which is animate, immortal and indestructible. If non-human animals can do without believing in God or any mystic concept why cannot we, humans? A cow or a dog does not seem to have any faith in God. Their life and activities are not in any way influenced by their such non-believing in God. Since man is endowed with superior level of intelligence and sense we venture into explaining each event and fact of life. But there are certain limitations to man's capacity. Combined with this the ego factor plays a dominant role in shaping his mental framework. Man's pride in his superiority is so unbound and capacity so limited that it is impossible for him to accept that he cannot understand the mysteries of this world and its events. In his overpowering and all—embracing ego he cannot also accept that any other human individual can have the power superior to his. So it is in all likelihood that he slips into making some foray into an unknown gigantic mysterious and inaccessible being that only has the potency to create and destroy, regulate and control, know and dispense his life and the earth he lives upon.

One man cannot accept that any other member of his species with the same structure as his is and with the same torso can possess supreme knowledge, power or truthfulness. This issues forth from his vanity. The entire edifice and structure of theology is conceived, created, nurtured, sustained, developed and carried forward in our mind only. It draws its vigor and strength in the fertile soil of human brain. It gets its nourishment from the brittle and palpable heart. It beams its exuberance towards the firmament of civilization from our thoughts only. It has no other abode or address. It starts in our mind and ends there.

In Upanishad God is stated to *be existing in and outside us, He is far and near, He is within everything and without everything.* These are postulates only without any substantiation anticipating an abject surrender to it by all. And he, who for any reason, raises doubt or asks for clarification is furnished with a package of stereo-typed explanations which further compound the confusion. If he is keen on better elucidation he will be dubbed as non-believers or infidels. It is common to all religions. In

17

some religions the doctrines are so rigid that injunctions are issued by their religious heads to behead those who dare to challenge the authority of the texts or the authority of the priests issuing such injunctions.

God is an abstract concept of perfect manhood which entire mankind strives to reach. Perfect manhood, in any way an idealism, when all the vices vanish or reach zero value is as impossible to attain as it is, in science, to create perfect gas or ideal gas where the molecules can move freely in any direction without collision or impossible it is to reach absolute zero temperature when the volume of matter is zero (which is absurd).

For that reason we cannot discard the idea itself. We have to cling on to the idea simply because we need it.

CHAPTER 3

Nirvana

Death takes its toll on every living being. Great or humble nobody is spared the killing spree of death. Valiant warriors, eminent scholars, connoisseurs of art and literature, great genii of science and commerce are born in every age and country but once they leave this world few people remember them and fewer still of their achievement or contributions are recognized by the posterity. From across the distant shores of time the only face that surfaces is that of Buddha, the enlightened one. The face, one beaming with hidden glow of smile slightly bent meditating eyes casts a spell of serenity of an aura unsurpassed by any one.

Social Background: The age in which he was born and grew up was in great turmoil. There were unending squabbles among races and cults and blood-bath amongst the power hungry chieftains. The society was saddled with the burden of castes, cultural life became non-existent, and people followed ceremonial and ritual part of religion shorn of philosophical imports and human appeal. Many people were disgusted with the external formalities of Vedic precepts. This angst culminated in a new school of thought who questioned the authority of the Vedas and its decaying rituals and dogmas. Atheistic Sankhya, skeptics like Sanjay Belathiputta, materialist Ajita Kesakambali greatly influenced Buddha. The literature which grew out of this movement produced the idea of transcendental freedom (*nirvana—i.e. the freedom from self*).

Attainment of Nirvana: Buddha realized in the course of deep meditation (which is nothing but temporarily withdrawing from temporal world) certain enlightening truths about life and the world. He gave up the comforts of princely life and became a wandering ascetic. In this period he attained from his guru and a renowned sage, the Alora Kalama, the 'sphere of no-thing' which is very high mystical state. But Gautama was not content with this concept of no-thing. He longed for absolute truth.

Next he went to Uddaka Ramaputta who taught him to attain 'sphere of neither-perception-nor-nonperception, a higher mystical state than the former. He was not satisfied this time also and left that place. Finally he reached a place near Gaya. He started a severely austere life, practiced various self-mortification. As a result of which austerities his limbs became like withered creepers with knotted joints, hairs rotted at the roots fell away from his body. His famished body and weak health made him unconscious almost to the point of death. After having recovered from this moribund state he realized that this mortification will lead him nowhere. He changed his way of life and started taking proper amount of food and leading a normal life. His companions left him. He did not give up his quest for truth. One evening under a Pipal tree he sat cross—legged for uninterrupted meditation.

During the first part of the night he gained the knowledge of his former existences. During the second part he attained the 'super-human divine eye', the power to see the passing away and rebirth of beings. In the last part of the night he gained the knowledge to overcome all the cankers and defilements and the four noble truths. His mind was emancipated, ignorance was dispelled, knowledge arose, darkness was dispelled, and light arose.

Concept of Nirvana: According to Buddha: 1) Existence is painful. 2) Individuals are handicapped by limitation which gives rise to desires which in turn leads to suffering since the desired object or subject is impermanent and indefinable. This temporariness causes disappointment and '*dukkho*' or sorrow. 3) This sorrow can be eliminated/ overcome. 4) This is possible if individuals follow a path shown by him. Now what is that path?

We have to examine certain terms used by Buddha and correlate them.

Soul—it is not a metaphysical substance. It is the subject of action in a practical and moral sense. Life is a stream of becoming (remember Tagore's definition of God as man's journey towards becoming Man).

Self—People prides on his own fortune, wealth, fame, social status, family lineage. But these are not permanent. Hence there is no own or self.

Human being—as individual has no value; human existence is the composite of five aggregates—i) form (*rupa*) ii) feelings(*vedana*) iii) ideation (*sanna*) iv) mental formations or dispositions(*samkhara*) v) consciousness (*vinnana*) none of which is self or soul. A person is in a continuous process of change with no fixed underlying entity.

Karma (action)—Good conduct results in happy ending and prompts further good acts and evil conduct brings in evil with repetitive tendency. Some karmas yield fruit in the same life, others in the immediately succeeding life while others bring fruit in far remote lives.

Here lies the paradox. On one hand he refuses to accept self or ego on the other hand he believes in rebirth. How is rebirth possible without a permanent self? He considers that human being will continue to suffer from sorrow until he overcomes his craving for desire and will be born in this Sansara again and again. This cycle of rebirth will end only with the extinction of fire of passion, desire and lust.

Eightfold path for good conduct—right view, right speech, right aspiration, right conduct, right livelihood, right effort, right mindfulness, right meditational attainment. How can we define right?-only a right person can conceive of this righteousness.

Who will determine who is right or wrong?

Nirvana—The goal of religious precepts according to Buddha is to liberate man from the bondage of ego (in Latin ego means I) which will relieve him from the shackles of this mundane world. One who succeeds in overcoming the pressure of ego will be freed from the chain of rebirth and with that he will succeed to win all his dukkho or sorrow. This is the state of Nirvana or enlightenment.

Literally Nirvana means dying or dousing the flames of fire, fire of lust, anger and delusion. But Nirvana is not extinction. Craving for avoiding sorrow is also a fetter. Salvation not cessation is the ultimate goal. But what will happen to the person who has attained Nirvana? Buddha avoided this tricky question by saying that it cannot be answered from within the confines of ordinary existence. What instead he asserted that this can be experienced in this present existence also. He did not categorically explain what lies beyond this phenomenal world.

The state of Nirvana can at most be accepted as a process of tryst towards a mental condition when one can get rid of the causes of sorrow i.e. craving for desire, lust etc. Again we come across a principle or an idealization of a concept that endeavors to liberate man from the pangs of death, disease, senility and non-fulfillment of expectations through a control process to form disciplined mental faculty in which ego or self is dissolved and impatience for not having achieved the targeted goals does not mar the equanimity. Having thus set a goal for salvation from pain he explained that there is something unborn, unoriginated otherwise man could get out from the pains of being born.

In the first place the idea of rebirth is in itself silly from the view point of biological science. We take note of two points i) Realisation of nirvana is possible in this life. ii) This is not possible to answer what nirvana is within the confines of ordinary human existence. Precisely Buddha indicated that the unending cycle of rebirth causing continuous suffering growing out of desire is not the central theme otherwise he would not have told that this is possible to achieve nirvana in this life. But this is achievable only by extraordinary humans. Such esoteric knowledge is of little value for this world is constituted by numerous ordinary human beings. If a religious teaching does not benefit the common mass its relevance comes to a naught.

We have discussed the complex and huge structure of Buddhist religion in a succinct manner. For our discussion this is sufficient. Without getting into the knitty—gritty of its various branches and its development in different countries and different ages we can have a comfortable grasp from the texts and related excerpts. Thus we know that despite variations in external formalities and interpretations all Buddhists follow the basic concepts as presented above.

Buddha repudiated the idea of ego or self on the ground of its being impermanent. However ruinous the effect of ego we cannot afford the luxury of ignoring all-pervasive influence of ego in human life. In psychoanalytical term ego is self or 'I' experienced with the external world which is understood through perception. A portion of human personality interacts with the external world, physical and social. When a man is born and grows up his senses grow and with this ego develops. (Without delving into the philosophical intricacies we will use the term in its usual connotation as the common man understands).

It is the natural yearning of a human being to seek his self-interest and profit for own. The ego is spontaneous and induced. The latter is built upon the influence primarily of parents, his lineage, society, group, geographical attachments, historical memories and a common cause which bind together a number of people. Ultimately the individual self seeks mental satisfaction through fulfillment of his own desire or the desire shared by him along with others. The common bond represents a religious, ethnic or linguistic affiliation.

But for the self-assertive ego the human society could not have developed. The massive structure of human civilization is built upon the strong driving force of self or ego. We are what we are only because every human being, noble or ignoble whatever way we may interprete, possesses this in-built urge of self to dominate others as an individual or as a group.

This manifestation of self is evident in small groups also. When for example in a football match or cricket tournament one team lobs a ball into the net of the opponent or takes a wicket the former bursts into frenzied joy while the other dips into gloom. Does it not seem to be cruel that one man or one group of men (in this case the players and supporters of the team) rejoice at the expense of sorrow or pain of another? Here is an apparent example of gratification of one's self or ego in exchange of a proportional amount of anguish of the other. In the same breath it can be stated that without this reactionary nature of self-satisfaction no man will have the gusto for performance or keen zest in delivering the best of one's capability. This is a universal phenomenon which cannot be refuted or debunked by any stretch of complicated argumentation.

Relevance of Buddha: Against this back-drop Buddha's idea of dissolution of self leading to nirvana seems to be naïve and palpable. We have seen in our life past and present the evil effects of ego unbound and uncontrolled. Buddha's sermon if taken in a positive sense guides us in this dark maze of life as a beacon of light like the pole star. From far above us he stands alone to say 'Men, please hold your desires and lusts under control. They are like fires which unless extinguished shall gobble the good things in life.' He of course was wise enough to understand that this fire cannot be doused totally but simultaneously he contemplated and felt the imperative need that to salvage mankind from the pangs of disease and death, from the clutches of gloom and despondency this idea also must be percolated amongst all that our desires have to be capped if not maimed altogether. Nirvana is only a philosophical theorization of this concept. We need not read more into it.

Buddha was told by his disciples that the Vedas prescribes for differentiation of man on the basis of caste.* He advised them to ignore Veda. Asked about the existence of God he admonished the questioner and counseled him to desist from being inquisitive about God and to concentrate on more pressing problems of common people and address them. Thus a great rational humanist appeared on earth two centuries and a half back.

---

* Caste system initially started as a form of distinguishing men on the basis of their profession or vocation. This was necessitated as a means to ensure professional expertise, integrity and dedication. In this sense it was one of ancient economic classification. This activity—based categorization degenerated and came to be identified with one's parentage. A person born of a potter parents came to be identified as a potter and the like. The system became so much regimented that one caste refused to dine with a so-called lower caste according as four broad castes enunciated in Vedas (Brahmin, Kshatriya, Vaishya and Shudra) Inter-caste marriage and social mixing almost stopped though not fully prohibited.

CHAPTER 4

Dictatorship of Proletariat

Karl Marx, the great thinker of the nineteenth century who espoused the Dialectic Materialism* wrote, *"Religion is the last resort of the oppressed masses. It is the circle of light in the valley of sorrow."* He didn't support or assert the existence of God neither did he totally negate the idea of God or religion. He struck at the root of the issue by upholding the necessity of religion in strife-torn grief-stricken human life. Marx is the solitary individual in the entire history of mankind who has shaped the course of socio-political movements all over the world directly or indirectly. Marxist doctrine has been adopted by more than half the population of the world and the rest of the countries alarmed by the specter of Marxism have suitably adjusted their social objectives. But his theory to evolving an all-time solution to the capitalist regime and installing dictatorship of proletariat is a myth.

Genesis of Marxism: Nobody can, however dispute his statement that everywhere the capitalists used to exploit the labor class and unless checked by social upheaval that exploitation will continue unabated. Marxist doctrine is not a panacea to the exploitative feudal and capitalist forces. He deduced that only a tiny fragment of the world population usurp the major chunk of wealth of the earth while the rest remain in a depraved condition. They are ill-fed or unfed, ill-clothed or do not have cloth worth its name, live in most unhygienic conditions in shanties or ramshackle units. They live in a sub-human condition with no guarantee of life , suffer perpetual deprivation and humiliation , under constant threat of losing their paltry means of livelihood, the threat of loosing of which humble means force them to accept further exploitation that accelerate the pace of decay leading to grave earlier than they should occur. Up to this point he is perfectly alright.

Trouble starts and we stumble upon what he proposed as a way out to this knotty situation to liberate the impoverished people who have only their physique as their weapon of survival. The latter termed as proletariat shall replace the ruling elite backed by feudal lords and capitalist barons. Marx made class struggle as the central force of social evolution. He wrote, "*The history of all hitherto existing human society is the history of class struggle.*" Well, according to him two basic groups exist (Or, existed in his time?) besides other groups of lesser importance existing around them. And they are in constant struggle against each another. One, the owners of means of production or the *bourgeoisie* and the other, the labours or the *proletariat. The forces of production which develop in the midst of bourgeois society create at the same time the material condition for resolving this contradiction and with this the pre-history of human society ends. The bourgeoisie are its own grave—diggers. The fall of bourgeoisie and the rise of proletariat are inevitable.*

In his scheme of things a permanent revolution involving a temporary coalition between petty bourgeoisie and the proletariat rebelling against the capitalist is conceived. The proletariat continues its efforts alongside to capture power without the petty bourgeoisie. Once a majority is secured the proletariat will secretly assure power to itself against the bourgeois authority. The process of educating the masses including the proletariat and other subservient classes continues until the proletariat comes to the core of power. And install a dictatorial form of government with the proletariat at the helm.

Weakness of Marx's Idea: The whole scheme sounds inconsistent and too simplistic. Marx lost sight of the fact that human society did not originate and does not evolve on the basis of any conscious mathematical efforts. In his scheme he missed to include the most complex, elusive, indefinite and indefinable element i.e. the human mind. The continuously changing and unpredictable human psychology does not follow a rigid or static pattern. We can at most discern certain traits of human character. It is common in all human society down the ages that one man will dominate others or one group of men will dominate another (e.g., racial group, religious group or group based on monetary power etc.) The dictatorship thrives on this basic psychological feature. Hence quite apparently dictatorship in any form is vicious simply because of its own burden.

The dictatorship of proletariat is still more absurd; any form of government unless checked and counter-balanced by various organs speaking on behalf of the ever— vigilant citizenry will degenerate into a sloth and greedy organization where a group of people undaunted due to absence of censure for its lacunae wield enormous power.

The art of statecraft enjoins to have a constant vigil over a vast multitude of people scattered over a large geographical space and historical time. The pros and cons of running an administration command a sound knowledge base.

With a handful of people in power it is very difficult, nay impossible to keep track of necessary details for the governance. The powers gravitate round a small coterie of people who supply the information acting as the organs of the government. This group eager to enjoy the pleasure of dominance and handicapped by its limited capability mostly supplies wrong and erroneous facts and figures, malafide, concocted, half-baked truths, spread conceited half-truths and ultimately one man surrounded by self-seeking cohorts and sycophants grab absolute power and unleash a reign of terror almost always in the garb of a benevolent superman.

This concept of benevolent dictator is illusory and the result of a naïve and wishful dream. Once in a blue moon this may occur. But we cannot leave this concept to thrive and throw the fate of hundreds and thousands of people on the mercy of a small group of people forming a party or to a seemingly good person, to say the least. Now coming to the point of dictatorship of proletariat we can say this is inherently silly. Once a dictator the proletariat does not remain as such. Power corrupts and the proletariat is not an immobile rigid block not to be influenced by its ill-effects.

We want to establish two points-one Marx's genius in diagnosing the evils of an unchecked 'capitalist' society and at the same time ludicrous folly of that great man in prescribing the medicine to cure the disease. The social contradiction that Marx so poignantly elaborated crept into his own theory. His identification of the then existing social inequilibrium is of paramount importance in the evolution of a thought process that initiated a dialogue between the oppressor and the oppressed though entangled in militant posture. His credit lies in the fact that he was the first person in the history of mankind to detect the germ of social inequality. He went on to explain:

The capitalist advances funds to buy raw material to produce finished goods which he sells in the market at a price which is more than the advance value. This money he reinvests to produce more consumer goods. Thus the fund he invests is not only in circulation its value increases in a compounded way. This incremental value advance produces capital. The labour power is the amount of labour required to produce sufficient for his and his family's subsistence. The capitalist uses the labour to work more than what is considered adequate for his livelihood. Thus the labour is exploited for the benefit of the capitalist. This is, in a nutshell, Marxist doctrine of capitalist exploitation of labour.

We do not dispute this economic explanation. Marx's theory carried by Engels and put to practice by Lenin, that the proletariat will come to power and the state will wither away is misleading and entwined in contradiction. No country can run without a government. Man cannot come out of its basic animist instinct and at a higher plane his emotional outburst of anger, greed and lust will not give in and to tame a huge mass into a disciplined and regulated life the necessity of presence of a strong state machinery cannot be wished away.

Significance of Marxism: Thus the Marxist doctrine falters on the ground in more ways than one. Notwithstanding the demerits cited above, Marxism stands apart as the singular phenomenon in the annals of human civilization not only for its novel thinking but also for the fact that he symbolizes the combined voice of the oppressed human soul; it whips our conscience and sensitizes us that there existed a criminal disparity between the haves and the have-nots. Though riddled with inner contradictions the applied Marxism flourished in Russia , China, Cuba, East Europe, Vietnam and its partial but no less epoch-making presence in India has left an indelible mark in the evolution of a socio-economic-political system which ushered in an emancipated labour class.

The statistical figures evince a sea-change in the quality of life-style of the labour, peasantry and other toiling masses. The increase in rate of literacy, decline in infant death and death due to starvation, increase in average life span, inequality in income distribution, reduction in unemployment and increase in gross domestic product are the indices of development and empowerment of the masses. Communism has faded in many of the countries pioneering the movement and the Chinese experiment with combination of capitalism and Marxist Communism is under watchful gaze of the world.

It has left a question mark on the relevance of Marx's theory in the wake of a radical change in the capitalist and labour relation heralded by the introduction of Welfare state and various means of production. If the figures put forward by economic survey by journals are any indication the communist countries though now forsook communism to pave the way for constitutional democracy based on universal franchise they will never go back to the bad old days of pre-communist era.

Now the social equations have undergone a cerebral change. The period in which Marx lived witnessed a spurt in industrial activity and it was accompanied by a steadily growing yawning gap between the owners and the workers. The glimmering disparity gave birth to simmering discontent brewing in the womb of capitalist society. The need of the hour was to have a towering personality to show the path and Marx did exactly that.

When Marxism was applied it bore beneficial effects, at least in changing irreversibly the character and edge of social discrimination based on economic exploitation that lasted for substantial period. The proponent of free-market policy also realized that in their own interest the labor class has to be kept satisfied to their optimum level. The competition among the entrepreneurs does not remain at individual level as a number of individuals may join hands to wrest a higher profit against another group.

The labor force if adequately paid or state shares the responsibility for looking after certain vital needs of labors and other suffering masses, like health, primary education, and agricultural inputs etc. they will feel secured and will joyously participate in the economic activity which will ultimately lead to more production i.e. more profit for the capitalist . The enrichment of one section and pumping of money will trickle down to the lower levels.

The concept of Welfare state thus evolved in capitalist social system with more zeal and uniformity than that followed in socialist communist countries only to avert Marxist communism. Thus one set of principle gets vindicated by sheer opposition to it. The second part of Marxist theory defeats the very basic concept of its first part i.e. to do away with the exploitative power from the bourgeois capitalist. What Marx envisaged by the 'dictatorship of the proletariat' is to snatch the state machinery or the manipulative power of the capitalist and vest it with the labour and other exploited classes of which was born a vicious system of producing another new demonic class of proletariat-dictator-cum-power-monger that will devour its own parents.

CHAPTER 5

Truth Viewed From Different Angles

Man's expedition for truth is contemporaneous with its civilization. The wise men of Athens—Socrates to Aristotle or the ascetics inhabiting the jungles and forests of ancient India hunted with primeval experience the nature and extent of truth. Truth is an abstract concept gained through real object. We cannot conceive or conjecture anything that cannot be understood by any of our organs. Scientific truth or theory is validated through experiments that are not always direct yet the idea basically flows from our experience gained by our organs and those ideas are transmitted and deposited in our brain.

Truth thus culled has been the subject of heated discussions throughout the ages. One such discussion will be taken up here. In the second half of 1930 Einstein and Tagore met each other four times (or more) in Berlin and New York. Two of these conversations are available. The first one appeared in the New York Times in August 1930 and in 1931 as an appendix to Tagore's essay Religion of Man. They have great significance not only in the study of physics but in understanding the whole gamut of truth.

Brian Joseph, a Nobel-Laureate physicist remarked. "Tagore is, I think, saying that truth is a subtler concept than Einstein realizes". Ilia Prigogine another Nobel-Laureate has declared that "**the new physics of quantum theory and the uncertainty principle are more indicative of the world view of Tagore than that of Einstein.**"

Herein below we reproduce excerpts from the meetings published in New York Times.

Tagore: You have been busy hunting down with mathematics the two ancient entities Time and Space while I have been lecturing in this country on the eternal world of man, the universe of reality.

Einstein: Do you believe in the divine isolated from the world?

Tagore: Not isolated. The infinite personality of man comprehends the universe. There cannot be anything that cannot be subsumed by the human personality, and this proves that the truth of the universe is human truth.

Einstein: There are two different conceptions about the nature of the universe.—the world as a unity dependent on humanity and the world as reality independent of the human factor.

(Tagore was deeply ingrained in non-dualism while Einstein's religion was dualistic with a God who can interact with the universe with or without man).

Tagore: When our universe is in harmony with man, the eternal, we know it as truth, we feel as beauty.

Einstein: This is purely a human conception of the universe.

Tagore: The world is human world—the scientific view of it is also that of the scientific man. Therefore, the world apart from us does not exist; it is a relative. world and depending for its reality upon our consciousness. There is some standard of reason and enjoyment, which gives it truth, the standard of the eternal man whose experiences are made possible through our experiences.

Einstein: This is a realisation of the human entity.

Tagore: Yes, the eternal entity.

Einstein: Truth, then, or, beauty is not independent of man?

Tagore: No, I do not say so.

Einstein: If there were no human beings any, then Apollo Belvedere no longer would be beautiful?

Tagore: No.

Einstein: I agree with this conception of beauty, but not with regard to truth.

Tagore: Why not? Truth is realized through men.
     After a long pause Einstein spoke very softly,

Einstein: I cannot prove my conception is right but that is my religion.

Tagore: Beauty is in the ideal of perfect harmony, which is in the universal being; truth is the perfect comprehension of the universal mind. We individuals approach it through our own mistakes and blunders through accumulated experience, through our illumined consciousness.

Einstein: I cannot prove but I believe in the Pythagorean argument, that the truth is independent of human beings. It is the problem of logic of continuity.

Tagore: Truth which is the one with the universal being, must be essentially human; otherwise whatever we individuals realize as true, never can be called truth. At least, the truth which is described as scientific and which only can be reached through the process of logic-in other words, by an organ of thought which is human. According to the Indian philosophy there is Brahman, the absolute truth, which cannot be conserved by the isolation of the human mind or described by words, but can be realised only by merging the individual in its infinity. But such a truth cannot belong to science. The nature of truth which we are discussing is an appearance; that is to say, what appears to be true to the human mind, and therefore, is human, be called Maya, or, Illusion.

Einstein: It is no illusion of the mind, but of the species.

Tagore: The species also belongs to a unity, to humanity. Therefore the entire human mind realizes truth; the Indian and European mind meet in a common realization.

Einstein: The word species is used in German for all human beings; as a matter of fact, even the apes and the frogs would belong to it. The problem is whether truth is independent of our consciousness.

Tagore: What we call truth lies in the rational harmony between the subjective and objective reality both of which belongs to the super-personal man.

Einstein: We do things with our mind, even in our everyday life, for which we are not responsible. The mind acknowledges realities outside of it, independent of it. For instance, nobody may be in the house, the table remains where it is.

Tagore: Yes, it remains outside the human mind, but not the universal mind. The table is that which is perceptible by some kind of consciousness we possess.

Einstein: If nobody were in the house the table would exist all the same, but it is already illegitimate from your point of view, because we cannot explain what it means, the table is there, independently of us. Our natural point of view in regard to the existence of truth apart from humanity cannot be explained or, proved but it is a belief which nobody can lack—not even the primitive beings. We attribute to truth a superhuman objectivity. It is indispensable for us—this reality, which is independent of our existence and our experience and our mind, though we cannot say what it means.

Tagore: In any case if there be any truth absolutely unrelated to humanity, then for us it is absolutely non-existing.

Einstein: Then I am more religious than you are!

Marianoff wrote."And with this Einstein exclaimed in triumph",

Tagore: My religion is in the reconciliation of the super-personal man, the universal spirit, in my own individual being.

Tagore and Einstein met in the house of Dr. Mendel in 1930 in Berlin.

Tagore: I was discussing with Dr. Mendel today the new mathematical discoveries, which tell us that in the realm of infinitesimal atoms chance has its play; the drama of existence is not absolutely pre-destined in character.

Einstein: The fact that makes science tend towards this view do not say good-bye to causality One tries to understand in the higher plane how the order is. The order is there, where the big combine and guide experience, but in the minute elements this order is not perceptible.

Tagore: Thus duality is in the depths of existence, the contradiction of free impulse and the directive will which works upon it and evolves an orderly scheme of things.

Einstein: Modern physics would not say they are contradictory. Clouds look as one from a distance but if you see them nearby, they show them as discrete drops of water.

Tagore: It is the constant harmony of chance and determination which makes it eternally new and living

Einstein: I believe whatever we do or live for has its causality; it is good we cannot see through it.

An incisive conclusion has been made by Prigogine who says, "**Einstein emphasized that science had to be independent of the existence of any observer. On the contrary Tagore maintained that even if absolute truth could exist it would be inaccessible to human mind. Curiously enough the present evolution of science is running in the direction stated by the great poet.**"

Einstein believed in the existence of a world independent of any observer. He thought that the moon was there whether one looked at it or not. And according to his idea it is the same in the sub-atomic world. This brought him into a direct conflict with quantum theorists such as Werner Heisenberg and Neils Bohr. Neils Bohr believed that it was wrong to believe that the task of physics is to find how nature *is*. Rather physics concerns what *we can say* about nature.

This debate with Einstein began in 1926 when Tagore first met Einstein and continued till Einstein's death in 1955. In the *Religion of Man* Tagore wrote" We can never go beyond in all that we know and feel.

Now let us take a look , very briefly at the two most engaging theories in this context; **i) Bohr's Atomic Model and ii) Heisenberg's Uncertainty principle**

(i) Bohr postulated that the angular momentum can take only those values equal to Planck's constant h multiplied by an integer and divided by 2Π. The angular momentum of an electron orbiting a nucleus depends on the radius of the electron's orbit, so this quantization implies that only certain orbits are allowed. The electron does not radiate continuously but it emits energy only when it moves from one fixed orbit to another. **But in this theory nothing is stated about the condition of the electron in the intermediate positions and how much time is taken in so doing**.

(ii) Uncertainty Principle states that certain pairs of physical quantities, in particular, position/momentum and energy/time can never both be known exactly. Indeed the better one is measured, the more uncertain the other becomes.

To measure the location of an electron we use a powerful microscope that uses a beam of very short wavelength, gamma rays. According to Quantum theory the beam consists of a stream of photons which can reveal the electron by bouncing it up the microscope. But each photon that bounces off the electron will change its momentum, just as in a billiard board. Indeed **the more accurately we try to pinpoint the electron by using the gamma rays of shorter and shorter wavelength the more momentum the photon gives to the electron because their energy increases with decreasing wavelength.**

The modern physicists have abandoned Einstein's view of sub-atomic world and they have taken a position that has close resemblance to that of Tagore.

CHAPTER 6

Ahimsa or Non-Violence

Man appeared on earth at the last leg of evolution. Yet he has dominated the earth over all other animals however furious it may be. This single fact clearly indicates that man is the most ferocious animal on earth. In fact man is the only armed predator of the world. Man's unique physical structure with a far more intelligence level than the cleverest of other animals have gifted him with a rare opportunity to conquer its adversaries. The superb intelligence combined with baser animal instincts made it possible for him to devise cruel, cunning and beguiling tactics to crush the opponents, first the other species and almost concomitantly the rival Homo sapiens.

Other animals can only utilise the weapons given to it by nature—teeth and claws, sting and poison. Man apart from these physical weapons created with its intelligence deadly weapons. In the primitive ages in the dawn of its appearance they used blunt stone blocks; first they crushed the supposed enemy by throwing boulders on it and crunched raw flesh. With the invention of fire they burnt the slain enemy or burnt it alive. With progress of time sharper and more effective weapons were added to its armoury. The blunt stones were sharpened to lend it better killing capacity. The process of manufacturing metals from crude ore was within man's reach. More and more refined shrapnel were invented—spears, arrows, swords, tridents, shields came. They were followed by fire arms— rifle, canon, guns and pistol; next came the chemical and biological weapons which can cause genocide of a horrific proportion; the last and most devastating weapon is the nuclear bomb.

Nuclear Holocaust

The world woke up to its horror when the first atom bomb was dropped on Hiroshima the sixth of August, 1945 that killed about three lakhs people and crippled ten times that number. Many theories have been advanced to rationalise this most barbaric assault on human race. The second atom bomb was dropped on 9th August 1945 over Nagasaki. The time chosen was 9-10 A.M.when people from outside the nearby suburbs flock there for various purposes so that the largest number of casualties can take place. Hiroshima is surrounded on all sides by hills and mountains. The reverberation caused by the abrupt release of huge amount of heat with consequent rise in temperature and fall in air pressure resulting in gust of wind and sound multiplies the destructive power hundred times.

It was argued that the Second World War could not be brought to an end had the two atom bombs not been used. However scurrilous that may sound this theory cannot be altogether refuted. The most deadly armoured and monstruous nations contemplated that another power stronger than they had appeared on the world theatre.

Germany, Italy and Japan came to their senses after recalculating their strength vis-a-vis that of United States, the new super-power. Such was the horror generated that on 10th August the Japanese offer came in the form of a meek surrender to the allied forces on condition of retaining the status of the Emperor intact and not surrendering the main island . The stronger elements in the junta batted for keeping the home islands. Finally the saner ones prevailed upon and on 14th august 1945 the Japanese army surrendered to the allied forces with assurance that the monarchy would be untouched.

*   *   *

French Experience

On 1st March 1716 Louis XIV the French emperor was murdered by Damion and was nabbed by the royal guard. The summary trial was held the next day and the verdict was passed. He was found guilty of regicide and was awarded corporal punishment. On 1st April, 1716 he was dragged on to an open cart drawn by mules. He had noting on except a shirt. A burning candle weighing about a few pounds was handed to him. The cart passed through open streets with on-lookers and passersby dotting the pavements on both sides and on roof—tops. He was standing on the cart holding the huge burning candle erect. Molten wax was rolling down his fingers but he could not murmur a protest. From the royal prison he was brought to the open terrace. Then his neck was fixed to the gallows with face upward and his hands and legs were tied with ropes to wooden poles. With prongs flesh was scooped out from his arms, thighs and chest and in the cavity was poured sulpher, acid and burning charcoal. Finally his two hands and legs were torn apart by pulling with ropes by horses racing in four diametrically opposite corners. The rest of his body was torched

The man was known to be very short-tempered and spewed slang interminably. Yet all these while he did not utter a single word but peculiar sounds came out from his quivering lips. A large number of people were invited to view this grand royal opera. Babies in arms hid face in their mother's breast in awe. And the Church priests were praying to God. It is left to the judgement of the reader to assess the rationality of this blood-curling cruelty. No amount of crime can be avenged in this way. Even the most dreadful of criminals deserve some sort of human treatment. But our wishes have little value when viewed in the backdrop of history of mankind.(The story was gathered from (History of Europe before French Revolution)

*　　*　　*

All Time Greats

Ashoka is presented as the greatest Indian emperor. Before Kalinga war he carried on wanton destruction of human habitations, plundering and blood-bath in all conceivable means. First he assured his authority over almost all of the Indian Territory. Thereafter he shunned violence and preached non-violence.

Alexander, the Great travelled all the way from far away Macedonia to the North West India upto Taxila (now in Pakistan). He is duly credited with discovering the road route from Europe to India thereby heralding a new era in the trans-national trade and commerce and social intercourse. He was embodiment of a real life legend. But his ascendancy was strewn with blood. He carried fire and sword wherever he went. In 336 BC the prince of Lyncestis allegedly murdered his father and he, in revenge, executed not only the assassin but all possible rivals and whole of the faction opposed to him. He crushed Tribally dispersed Getae, shattered a coalition of Illyrians. In the meantime with the spread of rumour of his death Theban democrats rose in revolt. Alexander headed for Thebes who refused to surrender. He penetrated and razed their city to the ground. 6000 were killed and all survivors were sold as slaves. The other Greek states were cowered by this ghastliness. In July 332 BC he stormed Tyre and achieved greatest military success but with great carnage and the sale of women and children.

Facts above marshaled from different sources are only indicative of true nature of human character in the course of history. Things or events painted or tinged with colour pervades our history books and collective memory only to make them acceptable to posterity. Had it not been so, the ugly face of human nature would be too difficult to digest for our pretence of civilization.

*   *   *

Modern India
Advent of Gandhi

Mohandas Karamchand Gandhi appeared on the stage of Indian National movement in 1914. He is credited with leading the freedom struggle from the front. But long before him the country had witnessed the people's movement. Initially it was confined to educated elite, landed aristocracies and middle-class aspirants to clinch equal status with their European or more specifically British counterparts. The seed-bed was sown long back with Rammohan Roy.

ROLE OF RAMMOHAN IN RESURGENCE OF INDIA— The entire Asia was engulfed in ignorance and unscientific mould of mindset in the second half of the eighteenth century when Rammohan Roy was born in 1772 (or 1774 according to another version). Poverty, illiteracy were the order of the day. Countries were administered by despotic extravagant rulers and society could not come out of primitive stage of savagery and barbarism. This man, Light of Asia and father of Modern India, hailed by historians as such, acted like a colossus. Single-handedly he carried on and fought with the British imperialists against vehement opposition of the indigenous orthodoxies to abolish *sati* the cruel ritual of burning alive the widow, especially younger ones, on the funeral pyre along with her demised husband. It was a mammoth task considered the social and cultural milieu prevailing then. It was a giant leap towards emancipation of women. In 1834 the *Abolition of Sati* Act was passed in British Parliament.

He met the Queen and pleaded with her to introduce modern branches of science—Physics, Chemistry, Mathematics, Biology, Economics, and Statistics et al. He strongly advocated the cause of setting up science laboratories. A historic incident took place in Carmichael Medical College of Calcutta in 1835. Dr Madhusuhan Gupta was the first person in India who made biological dissection of human body. The present generation shall scarcely believe the impetuosity of the epoch-making incident. It was considered a crime of grave nature punishable with social ostracisation.

We cannot overlook the imposing personality of Rammohan for this path-breaking deed. In political arena he was equally vocal for ushering in democratic set-up of government. His cardinal principal was initiation of liberal, rational, humanitarian, democratic and truly secular polity. India could not have been what it is but for the relentless effort of this lone crusader. If one goes through his interpretation of Upanishad it will be copiously clear that he had the best understanding of the underlying message of the *slokas*. With his incisively analytical style he established that the Upanishads adroitly proved the inanity of the fetish with rituals and associated mysterious and absurd beliefs and faiths. Amidst the mess of uncensored and banal insertions of the religious texts he picked the most relevant and well-connected knots in the *slokas* to derive at a conclusion that only fools lured by prospects of reward and cowered by fear of punishment engage in mindless celebration of ritual formalities and believing in stupid dogmas.

GANDHI'S ESTIMATE ABOUT RAMMOHAN—What was Gandhi's evaluation of this pioneer of Modern India? In 1921 he responded, when he was asked in a public meeting whether English education in India was a mixed evil, (his mulish, adamant and bigoted answer was printed in *Young India*, his own newspaper which indicates that this was not stated on the spur of the moment without application of mind; rather it can be construed as a well-orchestrated statement):

Tilak [the Bombay nationalist leader] and Rammohan would have been far greater men if they had not had the contagion of English learning [clapping] . . . I am opposed to mak[ing] a fetish of English education. I don't hate English education. When I want to destroy the Government, I don't want to destroy the English language but read English as an Indian nationalist would do. _Rammohan and Tilak . . . were so many pygmies_* who had no hold upon the people compared with Chaitanya [the founder of Vaishnavism], Sankara [the Hindu philosopher], Kabir [the medieval mystic poet and Nanak [the founder of Sikhism]. It is my conviction that if Rammohan and Tilak had not received this [English] education but had their natural training they would have done greater things like Chaitanya

But Gandhi made a volte-face in 1925 in tune with his characteristic mercurial nature; he wrote:

> "I have never anywhere described that great reformer [Rammohan] as a pygmy much less regarded him as such . . . I remember having said that he was a pygmy compared to the unknown authors say of the Upanishads. This is altogether different from looking upon [Rammohan] as a pygmy. I do not think meanly of Tennyson if I say that he was a pygmy before Milton or Shakespeare. I claim that I enhance the greatness of both. If I adore the Poet as he knows I do in spite of differences between us, I am not likely to disparage the greatness of the man who made the great reform movement of Bengal possible and of which the poet is one of the finest fruit."

Gandhi's appreciation of Rammohan was one of involuntary acquisition instead of being spontaneous assimilation. He was possibly quick to realize that he had

---

*   underlining is mine

made a hasty conclusion with the possibility of serious repercussions or some other person might have sensitized him about the demeaning assessment of the tallest personality of Indian resurgence. But one discerning mind cannot miss that he was still imbued with certain misconceptions. First he never properly grasped the role of Rammohan in the rebuilding of India. He placed Rammohan in same bracket with Tennyson vis-à-vis Chaitanya-Kabir-Nanak-Sankara with Shakespeare-Milton. This is glaringly true that he belittled Rammohan. Now the question is why he did so.

He was never a connoisseur of art and literature. Once he visited Ville Nauve, France to meet Romain Rolland in March, 1932(?) along with Mahadev Desai and Pyarelal; he rebuked them as soon as they initiated discussion on literature. He even prohibited them from indulging in any cult of intellect and engaging in art or literature. (*'Inde1915-1943'* by Rolland). Krishna Datta and Andrew Robinson have commented *'In Rabindranath Tagore—the myriad minded man'*—"*. . . . Romain Rolland who admired Gandhi without understanding him . . . . Gandhi seems not to have acquainted himself with even Tagore's best essays in English translation-let alone his many short stories. Either that, or he was pretending publicly that Tagore was something he privately knew him not to be. Probably there was both ingenuous ignorance and ingenious insincerity in the Mahatma's response. Gandhi was never much of a reader of literature.*" Gandhi wrote *'If the Poet (Rabindranath) span half an hour daily, his poetry would gain in richness. For it would then represent the poor man's wants and woes in a more forcible manner than now.*"

Who will explain to him the essence of poetry not being related to directly and dependent only on harping on poor man's wants and woes? This issue will be discussed at full length later on. Poetry or literature, to be more specific, does not deal with the daily routine life of common folk until and unless it is suffused with sublime realisation of a greater truth and general empathy about all the human being, the myriad intricacies of life and this creation with complexities of human mind in the one-to-one and one-to-many relation and finally the ultimate purpose of life with stoic acceptance of vices and virtues of man. A man of literature is not a preacher of morality. The alleviation of poverty and amelioration of the downtrodden are the job of basically the rulers, political workers, economists and social scientists. A poet's job on the other hand is to compose and create a good piece of poem and nothing else. A single poem like Tagore's *Golden Boat* or Keat's *One Day in the Life of a City Pent* or Jibanananda's *Banalata Sen* speaks volumes that may not be covered by an epic or a lengthy prosaic composition. But Robinson writes ". . . . *it is hard to believe that by 1925 he (Gandhi) was not aware of Tagore as an acute and trenchant critic of the rich and powerful, with a concomitant deep sympathy for the common man and woman*".

Now it is given to understand that Gandhi intellectually did not belong to that category whence one can expect him to be rational and free of superstitious belief. On the other hand Rammohan was a product of the best of western civilization. And we must unequivocally admit that despite aberrations the West only has liberated the human race from its countless chains. The Indian philosophy bogged in quagmire of irrelevant and absurd mysticism in which till now so many so-called intellectuals in India and sympathizers of their view in the West thrive only helped perpetuate enslavement of the poor and have-nots.

These hapless creatures born as such had to live on their physical labor. The material progress does not necessarily mean death of spiritualism. It is rather the opposite. For survival of these very common people about whose welfare Gandhi seemed not to have a wink of a sleep had to labor fourteen hours a day to eke out a living. The condition of savagery continued till Western education, the best fruit of which is the modern science laid the path to reduction of working hours from fourteen to seven. The coming of machine age heralded by the discovery of motors riding on the success of modern science put an end to lengthy working hours because what could be produced manually earlier was being produced in less than half the time. This saved lots of working hours and people had sufficient time to engage in spiritual activity in real sense. Spiritualism is a kind of elevation of mental faculty by which man can turn the mundane daily life into a meaningful, ecstatic and joyous one.

*   *   *

SECULARISM AND GANDHI—Gandhi began his fast (and the last one) at 11-55 A. M. on 13 th January, 1948. New Delhi was swarming with waves of refugees hounded out of their homeland which overnight turned into a foreign land. It was carved out of India as a holy land i.e. Pak-stan or Pakistan. The idea of Pakistan itself is outrageous and betrays communalism to its core. We shall come back to it later on. Now let us look into the declared motive of his fast. The dreadful cold and misery at the refugee camps where tens of thousands of uprooted Sikhs and Hindus were huddled together drove these wretched men and women, the children and the old, the weak and the infirm, the sick and the dying and the impoverished and destitute people to occupy the mosques and homes deserted by Muslims. Gandhi wanted them to return these places to the Muslims and go back to the miserable camps. He also pressured the Government to return to Pakistan 550 million rupees and thus created a great divide in public opinion and Government of the nascent state. Whom he was trying to appease? The Muslim rapists and murderers?

Through his fast he fancies to appeal to God to purify the souls of all and make them the same. He even went to the extent of saying, a craven call '*Should all the Hindus and Sikhs in Pakistan be killed, the life of even a puny Muslim child in the country must be protected*'. His purported fanatical dream was to see '*All communities, all Indians to become again true Indians by replacing bestiality with humanity*'. '*If things*', he thought, '*cannot do so my living in this world is futile*'. Gandhi was, by all presumptions, supposed to have been acquainted with history of Islamist expansion.

Even by far-fetched imagination it cannot be held that Muslims, once they have scooped out a substantial portion of mainland India in order to extricate themselves from under the hypothetical fear of dominance of their greatest obstacle, the Hindus, to their campaign of converting India into an Islamist theocratic state and advance Pan-Islamist mission farther eastward, would ever return it to and merge with Indians.

In January, 1940 Mohammed Ali Jinnah declared that the Hindus and Muslims formed two separate nations who must share the governance of their common motherland. Three months later, in the Lahore session of the Muslim League (March, 1940), he declared that the Muslim must have a separate independent state. According to the famous Three Doctors (Dr. Majumdar, Dr. Raychowdhury and Dr. Datta) the idea had been first mooted by a group of young Muslims at the time of of the Round Table Conference but had found no support and was characterized by Muslim leaders as "*a students' scheme*", "*chimerical and impracticable*" *The great Sir Mohammed Iqbal, first in1930 and repeated in 1939 proposed for a loose federation of Pakistan comprising one or two Muslim states.*

The idea of Pakistan as a sovereign state received official seal of Muslim League at the behest of Jinnah in 1940. From then all attempts at reconciliation fell through. The British Government could also turn down the national demand through the medium of Congress Party for transfer of power as it would lead to violent backlash from Muslim bigots. The Congress did not accept Cabinet Mission's plan to form interim Government but agreed to participate in the Constituent assembly in order to frame the Constitution. The Muslim League reiterated that the *attainment of the complete sovereign Pakistan* remained unalterable. As the Viceroy refused to accede to League's (read Muslims') demand to go ahead with Cabinet Plan even without Congress and went back to his earlier plan to set up after the war a representative body to frame a new constitution to enlarge his Executive Council.

Jinnah repudiated the democratic system of Parliament government on the conception of a homogeneous nation and the method of counting heads as impossible in India. As an individual Jinnah was the least religious. He did not observe any Islamic rituals and never went to the mosques. But the combined Muslim psyche ensconced him as their masquerade. Jinnah's acumen led him to the conclusion that behind his apparent secular image he can deviously project him as the chief spokesperson of the Muslims. Nehru wrote, in *Discovery of India:*

Mr. M. A. Jinnah himself was more advanced than most of his colleagues of the Moslem League. Indeed he stood head and shoulders above them and had therefore become the indispensable leader. From public platforms he confessed great dissatisfaction with the opportunism, and sometimes even worse failings, of his colleagues. He knew well that a great part of the advanced, selfless, and courageous element among the Moslems had joined and worked with the Congress. And yet <u>some destiny or course of events had thrown him among the very people for whom he had no respect. He was their leader but he could only keep them together by becoming himself a prisoner to their ideologies. Not that he was an unwilling prisoner, so far as the ideologies were concerned, for despite his external modernism, he belonged to an older generation which was hardly aware of modern political thought or development of economies , which overshadowed the world today , he appeared to be entirely ignorant . . . he condemned both India's unity and democracy. 'They would not live', he has stated 'under any system of government that was based on the nonsensical notion of western democracy.' It took him a long time to realize that what he stood for throughout a fairly long life was nonsensical.</u>*

On the other hand though the Congress throughout its recorded history maintained secularism in its avowed principles and policies could not garner the support of the effective section of the Muslims. The role of Jinnah has to be re-written in the context of perpetrating heinous crime on the largest number of populace i. e. *Hindu Bengalis;* we shall take up that issue in course of time. The scale and extent of mass murder and untold miseries inflicted on these hapless creatures with Jinnah at the helm will ashame the dreaded and the most hateful historical personalities. Hitler and Musolini at least did not drive three

---

*   Underlining is mine.

hundred lakhs of people simply because they were born as Hindus.

As his conditions worsened Dr. Sushila Nayyar advised him to take proper

medicine, his naiveté touched its abysmal low. He said,' *If I have acetone in my urine it is because my faith in Rama is incomplete*'. Sushila replied, '**Rama has nothing to do with it' and explained the medical science associated with it** '. His reply that followed needs critical scrutiny. '**and does your science really know everything? Have you forgotten the Lord Krishna's words in the tenth chapter of the Gita—"I bear this world in an infinitely small part of my living."?**' It was almost a crazy fashion with him to say,'God will keep me alive if he needs' His mischievously cryptic remark about science exposed his illogical and superstitious mind. It is all the more serious that such silly statements emanated from no less a person than Gandhi whom the country rightly or wrongly considered their supreme leader. It has been confirmed convincingly that man's survival depends on factors that cannot be anyway related to any person, that too an imaginary epic character. And the whole world cannot be held in any part of a human body. It is a fabulous fantastic dream. Apart from some allegorical meaning it has no other value.

CHARKHA OR SPINNING WHEEL—Gandhi's obsession with Charkha was another example of his unscientific bent of mind. In 1924 Gandhi and his cohorts declared that all congress members must wear khadi at political and congress functions or while engaged in congress business, and each must contribute 2000 yards of evenly-spun yarn per month of his or her spinning or in case of illness, unwillingness or any such cause a similar quantity of yarn spun by any other person This would make India independent for her clothing and create a bond between the common people and the Congress.

To this Tagore wrote, '*One thing is certain, that the all-embracing poverty, which has overwhelmed our country cannot be removed by working with our hands to the neglect of science.*' He also wrote, '*To ask all the millions of our people to spin the charkha is as bad as making the easiest of offering s to Lord (Jagannath) . . . . There are many who assert and some who believe that swaraj can be attained by the Charkha; but I have yet to meet a person who has a clear idea of the process. This is why there is no discussion, but only quarrelling over the question . . . . If any true devotee of our motherland should be able to eradicate the poverty of one her villages he will have given permanent wealth to the thirty three crores of his countrymen.*' . . . . "*Carlyle may have proclaimed the dignity of labour in his stentorian accents but a still `louder cry has gone up from humanity age after age testifying to its indignity.*" Is it a likely policy to reassure our grown up people by telling them that they will get their swaraj—that is to say, get rid of all poverty, in spite of their social habits that are a perpetual impediment and mental habits proclaiming inertia of intellect and will be simply twirling away with their hands ? No! If we have to get rid of this poverty which is visible outside, it can only be done by rousing our inward forces of wisdom of fellowship and mutual trust which make for cooperation.

When the civilization is racing towards a production mode with less and less physical labour advocating for spinning wheel (charkha) that entails lots of energy and time is not only retrograde, it is economically and commercially wasteful luxury. A ship run with thousands of ores is much less viable than the one driven with sails

---

* *Underlining is mine.*

harnessing the mechanical energy of wind and a ship run with motor is further more labour-saving. A cotton—mill produces much larger quantity of yarn than any hand—driven apparatus. The exploitation of natural forces to its own benefit and thus making intelligent use of nature signals victory of man over nature's forces which is otherwise destructive. In 1925 Tagore wrote an essay, 'The Cult of Charkha'. It was a scathing attack on Charkha policy.

Krishna Datta and Andrew Robinson writes, "*reading the debate now Gandhi's reasoning appears debilitating beside Tagore's (No wonder Nehru came to agree more with Tagore as he grew older). In order to evade the forces of Tagore's argument Gandhi took refuge in most un-Mahatma like sarcasm and in what can only be described as economy with the truth on the crux of their disagreement: the true stature of Rammohun Roy, the great reformer and founder of the Brahmo Samaj.*" We have already discussed the issue above. The motive behind Charkha is that it required minimum effort to rally people to feign to do something for the country. No sacrifice either in terms of career or endangering life was required. Sitting in the cool comfort of home without any fear of being arrested, harassed, tortured, deportation, jailed and be hanged one can spin yarn and earn the reputation of being patriotic. Nehru vainly and vaguely attempted to justify Gandhi's obsession with cottage industry, hand-spinning and hand—yarning when they appeared to have only marginal benefits. After independence he laid emphasis on building huge industries like oil Exploration Corporation, mining and steel industries, super sized thermal power plants and gigantic hydroelectric projects under master plans for irrigation, power generation and controlling floods in the second five year plan. He called them the temples of modern civilization.

NON-VIOLENCE AND GANDHI—Gandhi could not enthuse people with any higher ideals than being non-violent. But this sort of hypocrisy was designed to serve a greater purpose. The British people and the whole world were slowly moving towards a tacit realization that one day or the other they have to leave their colonies. But they had invested huge sums of capital, though earned mostly through exploiting the colonies and a substantial number of their countrymen were engaged in various administrative and mercantile posts. If the terrorist movement were allowed to operate their men and material, life and property would be severely endangered. To institute an opposite form of movement would crush the militant organizations without letting them understand that and whip people's sentiment that they have got an avant-garde to take charge of their welfare and liberate them from the bondage of foreign rulers. On the other hand the colonial force would be able to extract as much wealth as possible out of the impoverished nation.

Europe the pioneer in all forms of modernity and the chief liberator of the human bondage is capable of molding the world view in their favour by virtue of their superior intellectual capability and money power. A piece of great literary creation or work of art, be it a film or painting outside the western world does not get recognition in their own country unless and until it secures encomium from the western critics. It is both true and untrue. They can identify the real talent or genius especially in the domain of finer aspects of art, literature and philosophy because the Asian nations by and large do not have that clairvoyance of idea in this regard. And this hands out a weapon to the western scholars and people to create and destroy opinions, make and unmake great men, do and undo historical facts. Thus was made Gandhi, the superhero of non-violence.

Einstein wrote after Gandhi's death. *Generations to come shall scarce believe that such a man in flesh and blood ever walked on this earth.* Einstein never visited India and had scarce knowledge about its geography and history. They were fed by the strong British media and the government and he gulped it. In more than one occasion he has declared his faith in Jewish religion and pronounced that he is proud of belonging to that faith in public lectures in Jewish gatherings. He was offered to be the first President of Israel. He declined after a prolonged thought not because he was against the concept of a homeland for Jews in and around Palestine, their original homeland. He did not agree simply because he thought himself unsuitable for that post as he was a scientist and he felt free in that vocation and withdrew himself from that political responsibility.

"*He went to The United States of America primarily to raise money for the Hebrew University then being built in Jerusalem. During the preceding decade Einstein had become increasingly conscious of his Jewish background; more particularly, he had become convinced of the need for a Jewish national home. The growing anti-Semitism in Germany certainly played a significant role in this new awareness. Einstein established as an important supporter of the Zionist cause. After World War II his old acquaintance and Chemist Weizmann, became President of the new state of Israel. On Weizmann's death in 1952 Einstein was invited to succeed him. He declined on the grounds of lack of aptitude and experience, but the offer gives some indication of his standing in the Jewish world by this time*" (The World of Science).

But he was not aware that Gandhi vehemently opposed the idea of creation of Israel as that would displease the Indian Muslims. He would otherwise have revised his condolence message. We have already shown that Rolland another admirer of Gandhi appreciated him without properly understanding him. Rolland, contrary to popular belief also avoided any analytical estimate of Gandhi. His biography of Gandhi and Ramakrishna are written in a state of trance and display some kind of bewildered amazement of the mystic India.

Violence is a form of expression of energy. All round the globe is scattered the massive evidence of application of energy. Even Gandhi himself stated that if he had to choose between cowardice and violence he will prefer the latter. In eighth part, 7th Chapter of *Chhandogya* Upanishad it is explained as such:

"Force or strength is superior to knowledge. One strong person can cower hundreds of knowledgeable and wise men. If a person is powerful he can be enterprising and thereby is enabled to serve his teachers (these explanations are to be received with a grain of salt considering the prevailing ideas of the ancient age in which these hymns were composed) *and sit near them, listen to them, contemplate, appreciate and work and attain specialised knowledge with that. This world is based on this force. It is only with force that the space and the earth, Mountains, deities and men, grass and the trees, wild beasts and insects, ants everything including the heaven survives. Hence worship force."*

Analysing its immediately preceding and succeeding parts it can be concluded that these portions are not inserted and are original. These chapters are amongst the best parts of Upanishad and the fruits of deep meditational thoughts. Life itself is an anathema to non-violence. The great philanthropist and thinker M.N. Roy adds . . ."
. . . *it (Gandhism) is a utopia; indeed, even worse; it is a palpable absurdity. Utopia has an ontological as well as a*

61

logical validity. It is constructed out of imagination, on the basis of the world of experience. The Gandhist doctrine of non-violence has no such connection. Therefore it is an absurdity. It cannot be practised, not even by the prophet himself. Gandhi does not preach moderation of violence. His doctrine is absolutist.

The Gandhist doctrine of non-violence being so palpably absurd, logically untenable, scientifically false, empirically impossible, pragmatically negative, it should not be taken as the criterion for judging the social significance of Gandhism. Indeed this pseudo-moral cult serves the purpose of concealing the reactionary, and therefore immoral, essence of Gandhism. It has been constructed for this purpose . . . . Even the most peaceful revolution does some harm to its opponents. Violence is committed of necessity. Operation of force is inherent in all physical and biological processes, and social evolution, after all, is a biological process; the soul also has natural history. Therefore, absolute non-violence would rule out all social progress which, as any physical or biological movement, represents operation of energy".

Gandhi had an advantage; he had the backing of a whole nation. The cultural roots embedded in their psyche lured them to believe in submission, without any application of intelligence, to an all-mighty. This further leads them to acceptance of any kind of absurd belief. The spirit of renaissance which preaches for freedom of each individual from unreasonableness was absent. The cult of rational thinking initiated by Rammohun was throttled. The obvious precipitates of this were meek acceptance of physical and psychological torture like self-effacement, fast, penance etc. Man degraded to such a beastly ignorance of things happening around him cannot be expected to seek freedom in letter and spirit because that will require him to question the prevailing authoritarian forces including that of Gandhi's doctrine that the prophet of non-violence himself would not brook. People idle away their life and proceed to burial ground or burning crematorium by safely closing their eyes and reposing their faith on an unknown omnipotent force, even to the local chieftains or an ascetic leader like Gandhi. Thus is born the godfather.

In a dictatorship the subjects suffer from a smugness that they are very happy under the regime while the authority that they so ardently admire slowly removes from under their feet the very freedom of life. All on a sudden they come to realize that they have nothing to anchor on. One must not be misled to think that we are singing in praise of violence. Want it or not, cherish it or not violence is there and it will not subside by any amount of recapitulation of the virtue of non-violence. Buddha and Christ long before him preached for non-violence (though with a difference in character). But any student of history knows the course of ferocious killings and mass carnage in the subsequent two thousand years and a half.

What do we find in Gita which Gandhi held to his bosom? One very intelligent person (Krishna, who is neither a God which is an impossibility nor a historical character; no trace of his existence is available. Entire epic is a creation of a poet based on his experience) is advising a less intelligent person (Arjun) not to be emotional rendering him palpable to annihilate all his rivals who were his near relatives. Crude machinations, fatal weapons, treachery, bribery, cajoling, hypocrisy-all ingredients of violence were applied indiscriminately. And this was rationalized in the name of 'Dharmayuddha' (War for religion). Indian epics suffer from this hypocrisy. All acts or misdeeds are justified in the name of fallacious arguments. Buddha, as it is evidenced in the stories of Jatakas do not exactly prescribe for an all-out non-violence. In Ramayana Ram committed such dastardly crimes that would abase the low level politician-turned—criminals of modern times. But Gandhi harped on establishing Ram Rajya which is nothing but an absolute fraud. Ram told Sita after restoring her, "I have not rescued you for yourself; you are contaminated by the touch of a male who is not your husband. I came to restore the glory of a king"

Now how far did Gandhi's non-violence policy work when applied in India's freedom struggle? The idea of non-cooperation is in itself violent from the attitudinal point of view. It obstructs the normal functioning of the opponent and does cause irreparable loss to it. Violence has two distinct aspects or hues—physical and psychological. In Gandhi's scheme the latter is totally ignored. In our daily life we commit psychological violence on and often. The effect of such violence though invisible is far-reaching and has its ramifications no less obnoxious than that caused by physical violence. We should not forget that the idea of non-cooperation and Swadeshi emanated from Anti-partition movement of Bengal. The terrorist movement was an integral part of this movement and was the fore-runner of struggle for freedom of the entire nation. Hence it cannot be claimed that non-cooperation movement was at all non-violent in form and character. In Chaura— Chaura, UP twenty three policemen were torched to death when the irate mob shouting slogans in the name of Gandhi set ablaze the entire police station. Gandhi had to hastily call off the movement. The Quit India movement that was the final agitation programme of congress—led by Gandhi was by no means non-violent. Many post offices, police stations and government properties were destroyed.

On 8th August, 1942, the All-India Congress Committee adopted a resolution in favour of starting a mass struggle on the widest possible scale. This is the famous Quit India Movement. On the morning of August 9, 1942 the Government came down on the entire populace. Leaders were arrested and The Congress was declared an outlawed organization. According to Three Doctors, " *as there was no definite organization and a complete lack of leadership violent riots and assaults and sporadic disorders , such as the cutting of telegraph and telephone lines , damaging railway tracks, stations etc.,occurred on a large scale in different parts of India. The Government again adopted strong measures of repression including firing from aeroplane; according to official estimates more than 60,000 people were arrested, 18,000 detained without trial, 940 killed, and 1,630 injured through police or military firing during the last five months of 1942.*"

We do not find a semblance of non-violence in this episode. If, somebody argues, Gandhi were not arrested, the events could have taken a different course. The efficacy of an ideology is vindicated if it can be applied under any circumstance. That nation is the most unfortunate if the presence of a superman is required to salvage it. The absence of Gandhi and decontrol of non-violent oaths taken in his presence only prove the futility of that doctrine. We are not advocating violence. The civilization enjoins on all of us to restrain our bestiality and be human. But the basic instincts can hardly be obliterated. They can at best be regulated.

It cannot be believed that Gandhi, a shrewd reader of mass psychology didn't understand the limitations of his doctrine. Here we draw a line between the public utterances and inner convictions. People tells lie and great people tell more lies simply because their utterances have far-reaching consequences that may turn the tides. The highest executives of a country start their day telling lies.

As majority of people are swayed by palpable emotions rather than logical inferences they eagerly await their icons to speak in tune with their wishes and dreams however fanciful they may be. An astute politician, Gandhi read into people's heart and relentlessly gurgled that rubbish to be greedily swallowed by gullible people. Whether would he, if given the charge of administering the country, have dismantled the army? Or, he would have abolished the police or any paramilitary forces? Or he would have maintained a non-violent police who will not weild lathi, fire tear-gas shells to disperse an irate unruly mob, open fire under duress to quell an impending dangerous move to thwart an attempt to destabilize normal public life?

Gandhi's View on Population Control

One American lady came to India as early as in 1940's to preach the use of condoms. Gandhi ridiculed and showed her the exit door. He referred to the example set by Sri Praksh Chandra Roy and his wife, the parents of Dr Bidhan Chandra Roy, who abdicated the normal conjugal life and took to celibacy when they were barely in their twenties. He cryptically remarked that we know how to control population. True to his spirit Indian populace, beloved children of the father of the nation has proved the truth of his conviction in the capacity of self-control of his countrymen. India can now boast of being honoured with the distinction of the most populous and densest country of the world. India is the seventh largest country of the world. China, three times its size has been able to contain population to as low as 120 crore whereas India's population is 130 crores. United States of America that is two and half times that of India has a population of 20 crores. I have heard Biharis, the state that has the highest birth rate, proudly proclaiming that if ten per cent of Biharis are educated they would conquer United States. Even the orthodox Muslim countries and the Communist countries have adopted stringent measures to curb staggering population and registered phenomenal success.

Lasting effects of Gandian Thoughts-

Under the influence and guidance of Rammohan and Hindu liberals like Keshab Chandra Sen and carried on by Young Bengal Movement India could have emerged as a forward-looking, liberal, rational and truly liberated nation. Gandhi's most abominable contribution to the nation's destiny was that he not only thwarted the endeavour hard-earned by stalwarts of resurgence he put it in reverse gear. India instead of being an emancipated nation entrapped itself in the centuries old stage of decadence and irrational, mindless gimmicks, half-truths and hypocrisy of epic proportion. The spurt of hysterical craze of people towards observance of religious rituals to the utter neglect of humanitarian and scientific values in the present day India is a sharp pointer to the blistering effect of dogmatic utterances of Gandhi. One ex-Prime Minister of India (another Gandhi by surname) is reported to have visited on the eve of election campaign one false Sadhu who used to sit on the branch of a tree and anybody willing to have his blessings will receive it by the touch of thumb of right foot on his forehead. He did receive it. Numerous Babas and Gurus sprout like mushroom and people sink to the bottom of idiocy by show of obsequiousness to them.

Father of the Nation?

Gandhi is called Father of the Nation not by any legislative amendment or motion passed in the highest representative body of Indian masses. Once the great Indian Subhash Chandra Bose called him 'Father of the Nation' (One of the mistakes committed by him). The irony is that the Indian national government has declined to accord any importance to this 'Prince of Patriots' i. e., Subhas Chandra Bose, as called by Gandhi, while they have retained the title conferred by that Prince on Gandhi on the spur of the moment. And interestingly another great Indian (incidentally he is also a Bengali) Tagore conferred the title of Mahatma (Great Soul) on Gandhi. Such was the apparent enticing influence of Gandhi that even great minds were taken off their feet. The man was honest but not simple as it appears, loveable but cannot be followed. We shall repeatedly come back to this persona in subsequent Parts.

If at all anybody can be called the father of Indian nation he is Raja Rammohan Roy. But where no other country in the world requires a father why should we? It represents a paternalistic society nurtured by our epics.

**In this article I have depended on and quoted from the writings of various authors to whom I owe my indebtedness.

CHAPTER 7

India marches Ahead

India is poised to take centre stage in the comity of nations as was envisaged by Swami Vivekananda more than a hundred years ago. Backed by its emerging army of educated workforce and nourished by vast natural resources and an ancient base of rich civilization India's presence is felt conspicuously in every field of activity of the world—producing creative literature, contributing to basic science and technological innovations, making good films , significantly participating in world trade and industry and so on and so forth. Amongst the Asian nations barring China and Japan few can rival or match us in so many varied areas of excellence. In sports arena also we are showing signs of formidable performance. Even a school student can present countless examples of India's success story.

We have developed a sound Election process thanks to the efforts of certain individuals. It has attained an impeccable perfection that many nascent states are emulating us and hiring our officials to conduct free and fair elections in their countries. Our judiciary and jurisprudence considered as one of the best in the world, despite accusations of procrastination in delivery of judgment, is pro-active in implementation of social order and ensuring political correctness. The long hand of Judiciary combined with regimented Executive machinery can hardly be overlooked for the fruition of our democracy. None can question that as a nation doomed to deprivation by centuries of subjugation and chaos we have surged forward towards modern liberal society. Here we must press the pause button. Have we? How far is our society free of evils and degradation? Let us introspect and have an overview of the psychological base of our society.

Problems of Indian History

The texture of our collective national character, if any, has been woven around certain mores. There was no concept of religion in India in its usual connotation before Buddha. We adopted and practiced some forms of our traditional way of life inherited from a murky past. This was not derived from any fixed and rigid dictates. Rather it flowed naturally from age old habits and traditions. The Vedas contributed in shaping our social psyche. But there was no conscious effort to form a religious sect and the influence of the Vedas was nominal and it touched only a few enlightened people. In major part of the land cult of local Gods prevailed. Many Indians worshipped animist faiths. People deified trees, tiers, stones etc.

It was only after the flourishing of Buddhism a section of people already settled in India-they might have migrated from outside few centuries back—and adopting Vedic injunctions realised their separate identity and in an effort to assert their distinctive cultures engaged in a long-drawn battle with the Buddhists. The march of Buddhism could be stalled due to the resistance offered by a group led by Brahmins and Kshatriyas mainly. By the turn of the fourth century AD during the Gupta era thetre was a renaissance among this group that reached the pinnacle of glory when Vikramaditya Chandragupta II ruled. This period is commonly hailed as the Golden Age of Hinduism.

But how can we accept that as the word Hindu was coined much later after the advent of Muslims? Historian Romilla Thapar once quipped that Vikramaditya would have been startled to know that he was a Hindu. In spite of all these we may not be wrong to affirm that this is Hinduism. The Hindu cult, faith or religion, whatever way one can describe it, as we find it today, is encompassed in a broad spectrum and can be traced back to the ancient Indian roots enriched by the Vedas, the two Epics, Upanishads and Puranas. It can be linked to the civilization that flourished during the Gupta period. It has no beginning. So it is called Sanatan Dharma. It has no code of conduct that is why it is so liberal; it embraces within its fold believers and non-believers, theists, atheists and even agnostics like Nehru and Satyajit Ray. We have stark materialists like Charvaks and serious thinkers in non—believers like Kapil.

A basic drawback in chronicling the Indian civilization is that Indians are shy of recording the routine life. We are rather lazy in diarizing our daily events. It was even more relevant when there was no press or publishers. It is too difficult to piece together the events culled from sources here and there in a convincing manner to arrive at a historically conclusive fact. For the entire ancient period we have to depend on the mythologies and ballads verbally passed on from one generation to another. Herein lies the root of multiple social problems. Most of the authentic narrators of Indian history are foreigners like Megasthenes, Fa-hien, Hiu-en Sung etc. The only exception is Chanakya also known as Kautilya (derived from the word kutil for his shrewd diplomacy) who fortunately for us bequeathed to us a very authentic document describing the social, cultural and political history of Mauryan era in graphic details. In Middle age Ibn Batuta, a Turkish traveler described the history of the Sultanate period. Abul Fazal also had his roots in a foreign land.

The lack of historical sense has caused a lot of troubles not only for historians and researchers it has created huge confusions in fostering a good social order. It is behind wrong concepts and fundamentals leading to social and religious conflicts. Indifference to adhere to facts and figures and tendency to lean on fiction, gospels and hearsays produces hyperboles, occults, myths and superstitions. We wholeheartedly believe whatever our parents /elders say. While doing so we forget that however respectable, they may be wrong and they are quite so often. One my friend once informed me that 'bibhuti '(sacred ash) falls off portrait of one particular 'Baba'. As I refused to accept he blurted in defence that it was no less a person than his father who has seen it. I counted that there might be sixty billion fathers and mothers in the world. If all of them are assumed to be correct and truthful in every matter we would have been gifted with a different world. We love and respect our parents because they are parents. How many of us have very knowledgeable, talented, wise and omniscient parents? We don't need it either.

Ram is an epic hero. But unlike him Greek epic heroes have not been turned into God. And if Ram were God how can he be born? Where is the evidence that he was born in a particular place? If it is accepted that an old Hindu temple had been converted into a mosque the agitation was not worth the social cost it bore. Ram has been called *Narchandrama* signifying that he combined all the strengths and weaknesses of man. For this reason his character is so vivid, real and enchanting. We cannot mystify him and diminish his epic grandeur to taper into one pale God-like creature.

*　　*　　*

Till now we have been able to establish two points. Our epics and religious texts contain lot of insertions by persons of lesser intellects. Secondly human relations especially those who are thick and thin of our blood need not necessarily be perfect. Unfortunately in absence of monitoring the religious scripts and ancient literature or any written material for that matter are loaded with such insertions which have gained more importance than the basic philosophical imports and poetic beauty. In the jungle of inconsequential projections we lose sight of consciousness and golden realization of life achieved by thinkers through their experience and observation with a touch of genius.

Ramifications of lack of objectivity of Indian History.

Caste System—The most disturbing and long lasting impact of these unauthorized insertions is the introduction and consolidation of caste system. This is not only unique in the world; it has produced a legacy that has given legitimacy to a belief which is basically inhuman and defies all logic and rationale. How can human being be considered high or low on the basis of birth or lineage? It does not stand the test of biological science or any other branch of human knowledge. The racial discrimination of man as it is found in anthropology is based upon physical features and does not discriminate one race being superior to other.

The discrimination is not only amongst the Brahmins and non-Brahmins. It percolates amongst the different orders of the same caste. The famous social scientist Dr Ashis Nandi informs that that there are 2000 varieties of Brahmins in India. One sect does not even dine with the other. The strife between Vokkaligas and Lingaet groups of Brahmins in Karnataka is well-known9 RK Hegre and HD Devegowda respectively belong to these two groups). In Tamil Nadu they have restored the social status of lower caste and somehow patched up a face saving solution through Dravida Munnetra Kazhagham movement.

Every political party exploits the caste divide to win election. In UP politics is clearly divided on caste lines. The rise of a dalit leader with a thumping majority initiates the victory of the downtrodden but when she makes Brahmin politicians line up to touch her feet and enjoys a kind of revenge that is also casteism in reverse order. Recently in UP when a young boy chose to marry a girl belonging to another caste the relatives of the girl killed both and laid the copses on the open road in full public view. Even the mother of the girl is stated to have told the reporters that others are a taught a lesson so that they dare not venture in inter-caste marriage.

The Reservation policy followed by the Government has ensured and built up confidence among the low caste by offering them opportunity to compete with the higher caste in unequal social system. It is a fact majority of the poorer people belong to lower caste. But as with the passing years those who have been elevated financially and are in an advanced stage have started looking down upon others of the same ilk and lagging behind them. Thus a privileged class is being created. The demand is doing the rounds that reservation should be based on income criteria. Who will do the job? A corrupt social system cannot ensure preparation of flawless list of low income.

*   *   *

Neglect of Women—It is estimated by historians that about 72 million people were killed during World II. Of these 25 million died in combat, 11 million were killed in the Nazi holocaust and another 20 million perished in war-introduced famine. Amongst this we present another figure. Demographers and economists estimate that today over a 100 million women have been killed globally by societies that prefer sons to daughters. Instead of guns, bombs and gas chambers the genocide against women is carried out by abortions, drowning, strangulations and nutritional and medical neglect. Unlike in a war where the enemy is the target, in this genocide it is the girl child's loved ones who kill her. Unwanted baby girls are killed after birth-smothered with a pillow or cloth, drowned, fed poisoned seeds or buried alive. The most popular form of femicide is now abortion with the help of an ultra sound detection of sex of the foetus.

A UNICEF report says **India kills almost 7000 girls per day by abortion**. India Today published a report in 1986 that in Tamil Nadu alone more than 600 girl children were killed in a year by putting salt in their mouth immediately after birth.

Persecution against women is manifest in all hues. An old lady was abandoned by her educated husband. The Supreme Court awarded her alimony; the highest legislature of the country changed the law to nullify the court's verdict to appease the fundamentalists.

*   *   *

Pseudo-secularism

A Hindu has an advantage. One can criticize its systems and practices without the fear of being accused of sacrilege and awarded death sentences or ostracization. Under the protective umbrella of broad Hindu society Hindu secularists, politicians or intellectuals all alike condemn whoever speaks in favour of Hindus even on right ground. On the other hand they look the other way at the most apparently communal activities or utterances of other religionists or tacitly support them. This adds fuel to fire and offer weapons in the hands of Hindu fanatics if any. Even the Hindu liberals, at least a sizeable section of them feel hurt and let down. The suppressed anger finds an outlet in the form of silent apathy or they even become rebellious inwardly. Thus when MF Hussain draws nude pictures of Hindu Goddess he is supported in the name of freedom of expression while Taslima Nasreen is hounded from her own country. She seeks refuge in a country of her choice that she loves and especially in a city (read Kolkata) that prides itself on being the most liberal and secular she is forced out. None of the demonstrators against her writings in Kolkata on 17th November 2007 have read or understood a single word of her writing but held the city at ransom and terrorized people for hours.

And her fault? She depicted in lucid prose the numerous incongruities in her society marshaled from facts encountered by her since her girlhood days. In 'My Girlhood Days' she has exposed the male chauvinism vis-à-vis persecution of women and illiberal practices prevalent in a primal society. She has not even spared her father and indicted her own behavior. In her novelette, _Lajja_ that is an account of facts presented in the form of a fiction she has serialized the events that followed in Bangladesh in the wake of demolition of a disputed shrine at Ayodhya and painted pictures of persecution of minority Hindus. We put pressure on Nepal, the only Hindu Kingdom in the world to transform into a secular democracy but keep quiet when secular Bangladesh declares herself as Islamic state. These ultra-secularists born as Hindu though are definitely doing a disservice to the cause of secularism.

Now it is proved that the disputed shrine was indisputably a temple converted into a mosque. The facts available in this respect are as under. A Hindu temple existed there since the days of Vikramaditya Chandragupta Maurya II. For about a millennium there was no problem. A few years before the irruption of Babur across the Western border one Muslim Imam made a connect with the temple priest and started occasional visit to the temple and exchange the religious and theological matters with the priest. It was really very exhilarating and could pave the way for assimilation of two different cultures and bring two apparently extremely opposite form of religious beliefs closer to each other. But history does not follow as we wish it should.

Ironically the priest died suddenly without keeping an heir-apparent. The Muslim saint or Darbesh stepped in and occupied the temple and started offering prayer in the Islamic way alongside the Hindu form of worshipping the idol. Things were going on smoothly and could have been so but for the intervention of Babur. He destroyed the superstructure and built in its place one with an Islamic architecture and a dome that is typical of a mosque. But at one corner the Hindus continued to offer pujas. During the British regime in 1849 the then District Collector placed the sanctum sanctorium of the shrine under lock and key because of communal tension that started brewing up by the time fomented by, as luck would have it, the advanced section of the Muslims and not the illiterate and ordinary Muslim masses who were so far untouched or ingenuously unaffected by any sense of communal divide. So far so good. In the meanwhile the Hindus initiated movement to restore their rights of possession. Hindus were very much within their rights to reclaim its custody of the sanctum sanctorium. The Hindus kept on offering pujas till then. As the movement reached its peak in 1986 the then prime minister Rajiv Gandhi opened the lock. Immediately the Muslim Personal Law Board filed a suit with The Allahabad high Court. It has been proved by archaechological excavation there did exist a Hindu temple and the mosque was built on Hindu shrine in utter violation of religious propriety.

Fundamentalism of any kind or faith has to be cried down. It is unfortunate that liberal forces among Muslims are silent for fear of being ostracized or even threatened with death.

*　　*　　*

True Spirit of Religious Texts

Every religion contains beautiful messages applicable to all. If we know the substances of eternal truths proclaimed in Upanishad, Quran, Bible or Granth Shahib we will not keep ourselves enchained to a set of rites which border on dogma. The incidence of Rup Kanwar a young widow aged 21 year burnt along with her husband has not faded from our memory. Still people glorify this inhuman practice.

Rammohun Roy wrote "To *follow these practices is only optional. Those who have no reliance on Shastra and those who take delight in the self-destruction of women may well wonder that we should oppose that suicide which is forbidden by all the Shastras and by every race of man.*" In proof he has quoted the text of Kathoponishad,-"faith *in God which leads to absorption is one thing and rites which have future fruition for their object, another. Each of these producing different consequences, hold out to man inducements to follow it. The man, who of these two chooses faith is blessed; and he, who for the sake of reward practices rites is dashed away from the enjoyment of eternal beatitude.*" And also the Mundokoponishad: "*Rites of which there are eighteen members are all perishable. He who considers them as the source of blessing shall undergo repeated transmigration and all those fools who, immersed in the foolish practices of rites, consider themselves to be wise and learned and are repeatedly subjected to birth, disease, death and other pains. When one blind man is guided by another, both subject themselves to all kinds of distress.*"

It is asserted in the Bhagwat Gita, the essence of all the Smritis , Itihasas that, " *all those ignorant persons who attach themselves to the words of the Vedas that convey promises of fruition , consider those falsely alluring passages as leading to real happiness and say that besides them there is no other reality. Agitated in their minds by these desires they believe the abodes of celestial gods to be the chief object and they devote themselves to those texts which treat of ceremonies and their fruits and entice by promise of enjoyment. Such people can have no real confidence in the Supreme Being*"

But why then do the Shastra prescribe rites as the means of attaining heavenly enjoyments? As men have various dispositions those, whose minds are enveloped in desire, passion and cupidity have no inclination for disinterested worship of the Supreme Being. If they had no Shastra of rewards they would at once throw aside all Shastra and would follow their severe inclinations. In order to restrain such persons from being led only by their inclinations, the Shastra prescribed various ceremonies but at the same moment expresses contempt for such gratifications. Had the Shastra not repeatedly reprobated both those actuated by desires and fruits desired by them all those texts might be considered as deceitful. According to Upanishad, *Knowledge and rites together offer themselves to every man. The wise considers which of these two is better and which the worse. By reflecting he becomes convinced of the superiority of the former, despises rites, takes refuge in knowledge, and the unlearned, for the sake of bodily gratifications has recourse to the performance of rites.*" The Bhagwat Gita says, "*The Vedas that treat of rites are for the sake of those who are possessed of desires. Therefore, O Arjuna! Do thou abstain from desires.*" Buddha also preached also preached to get rid of desires. He achieved his concept of Nirvana based on these precincts which means abstention from desires that does not subject man to rebirth leading to Nirvana or absorption.

The same principle may be extrapolated to other religions. But most astonishing though it may sound every religionist lays more stress on rites than on knowledge. Since rites of one religion differ from those of others a rash conclusion is drawn that every religion is different from the other. But long before Quran it was propagated by Upanishad that great divinity of Gods is one, in 'Mahaddevanamasuratwamekam'. Hence the unity of God was established in India long before the advent of Quran and formless god was worshipped even by the Aryans who settled in India. The idolatry and polytheistic beliefs are contribution of Mahayan Sect of Buddhist and especially Jains

To followers of Islam who as per the dictates of Quran persecute/ kill the idolaters, polytheists and non-believers in their last messiah , a question—how is it possible for God who has created all creatures including man (if we believe in God) who is supposed to be omniscient, kind and unbiased to set one group of people against another? Or, is it concoction of religionists? For them a quote from the Quran—"All human beings belong to one race." (Sura 2 Ayat 213); "Oppose all evil by good and your foes will become friends; only those who possess virtue of patience can do so'" (Sura 41 Ayat 45). In Upanishad's message "Shrinwantu Vishwey Amritasya Putrah! " we find a clarion call to all human being—irrespective of caste or creed we are immortal—a robust optimism that human race will not perish by any force. Man would indeed be in a poorer state if he is allured by hope of reward or cowered by fear of punishment after death.

CHAPTER 8

Islam and India

Islamic and Islamist—Islam started with the avowed principle of *'Conquering half the known world'* by persuasion or persecution. In India about 1, 00,000 Parsees and 24000 Jews lived who were evicted out of Persia and Jerusalem respectively during the destruction carried out in Babylonian war. There is a difference between being Islamic and Islamist. Being Islamic means to follow the tenets of Islam in their daily life and mainly confine its locus within the ambit of personal activities. But Islamist forces are those that vow to Islamize the entire world even at the cost of mass destruction and by employing all means, hook or crook. We have nothing to say against the Islamic persons. But certainly we have every right to speak against and take all sorts of measures to destroy the Islamist forces once and for all. In this task we have to take the support of truly Islamic populace before it is too late to let our mother earth perish by entangling in an unending and meaningless strife. Let us join hands with the Sufis, Ahmedias, Qurdis, Qadianis and the true secular forces amongst the Muslims. I had ample scope in the capacity of Government administrator to learn the crude method applied to cower and convert Hindus. On one hand the Hindu India being divided in so many fragments are dastardly apprehensive of militant and violent Muslims and appease them to the extreme. The Hindus though majority always tremble at the slightest prospect of incurring Muslim vengeance in its merciless grievousness and fall prey to their false promise and with the preconceived fear that in case they deny to oblige the Muslim's offer of courtship a very hard and cruel revenge shall visit them.

Manifestations of Islamists' Activities: The September 9, 2001 that razed the twin towers of New York established the horror of pan—Islamist terrorism. In conjunction with this we have to add the following:

Bamian Buddha statue was destroyed by the so-called Taliban, one version of Afghan Islamists; several thousands of Christians were torched to death in Indonesia. In Philippines Christians are under continuous attacks by the same terrorist forces. In Thailand the Buddhists and Hindus are facing the suicide squads of Islamist terrorists. In Sudan the Christians were being mercilessly beaten and murdered by these fanatic forces. Now at last it has been somehow solved by partitioning the small country in Muslim majority Northern Sudan and Christian South Sudan. Communists believe that all the ills of a society stem from capitalist exploitation of the proletariats and the panacea lies in overthrowing the capitalist dominated ruling class and installing a proletariat government where minor issues like casteism and communalism will vanish magically once the "scientific" Marxism is applied. It is a day-dream. How can they explain the eruption in Communist China of intense communal strife in Guang Dong Province? In Maldives the state religion is Islam. It means that a non-Muslim can be anytime charged with sacrilege on the slightest pretext of offending Quran and he can be killed by any Muslim. This murder will not be construed as a crime. The Hindus are second or third class citizen. One truly secular person feels baffled at the queer silence maintained by the secularists of our country on this glaring example of fundamentalism, Muslim fanaticism and inhuman barbarism. When so-called Babri Mosque was destroyed the veteran Marxist leader lamented that India's prestige has suffered irreparable loss before the world. But all these improprieties escaped his attention. Or, did he knowingly close his eyes?

Arab Spring-a Myth

Kanwal Sibal, the former foreign secretary of India writes, '*The Arab Spring hardly describes accurately the phenomenon that has affected this area. The ouster of dictators by the street was too quickly labeled as a grassroots democracy movement. The skeptics saw the Islamists stepping into the political breach created by social media activists. In Tunisia, the government with Islamists tendencies has come to power. In Egypt, the Muslim Brotherhood and the Salafists have won a majority in parliament. The public in these countries has preferred to express its political aspirations not through secular parties but Islamic one, with no ideological roots in democracy. In Libya, the up-rising against the dictator was tribal in nature with strong Islamic affiliations and not a movement for democracy. In Syria , too, the opposition to the Assad regime has less to do with democracy than a sectarian effort to create a situation that would invite foreign intervention on the Libyan pattern.*'

The idea of democracy does not go with Islam where it is indoctrinated that the religious head or *Caliph* is the supreme authority. And the latter acts as an agent of *Allah* who is *one* and the *omnipotent*. The external form notwithstanding the Islamist phobia throttles any kind of dissent with power of Mollahs. And the communists, especially in India cry hoarse over the secularism and mollycoddle Muslims. But Muslims hate the communist the most because they do not believe in God (at least theoretically and in its published policy). When Columbia collapsed the head priest of Mecca announced publicly, *'It has been cursed by Allah because the crew comprised one American Christian, an Indian Hindu and an Israeli Jew'*. This issue will be discussed later on.

Plight of Bengali Hindus

In a Muslim majority district of North Bengal where I had been posted as an Executive Magistrate everyday a sizeable number of Muslim youths came with a young Hindu woman to the Collectorate to make an affidavit for securing their marriage simply by that affidavit. It is only an oath signed before a Magistrate. These Hindu girls were a victim of their love—jihad. The method is simple. Either be lured by false promise or face the consequences— from being raped, houses torched, pelted with stones, continuous ridicule, removal of the basic easement rights and cruel and suffocating methods of sustained torture. Once they disagree to respond to their lustful advances the Hindu girl and his family would be ruined lock, stock and barrel.

Hundreds of Hindu girls are missing from Bengal from various districts in the border entangled in love—jihad. Poor

Hindu girls are entrapped in love by ISI-backed Jihadist and sent to Kashmir or Afghanistan for jihad against India. For more than four decades the process is on. It is not to be seen as an isolated attempt by an individual. The history of spread of Islam throughout ages bears ample testimony that Islam (that means surrender) has a conspicuous tradition since its emergence to broadcast its base to any corner of the world by various means. Before we set on to a detailed discussion on this very important but grossly neglected issue of grievous crime against Hindu Bengalis we shall make some foray into the historical stages through which the situation has come to such a pass.

Traces of Intolerance in Islam

The intolerance of other religion is the most intoxicated charm for the Muslims. First they annihilated the numerous religious tribes of Saudi Arabia and turned towards their immediate neighbor. The Jews of Palestine were systematically expelled from their fatherland and they spread in different parts of Europe. But there also they were not accepted as the son of the soil. And were looked down upon by the original Christian inhabitants. These Jewish refugees by dint of their intelligence and labour could swim up the stream and overcome the hurdles and made a niche for them in the hostile alien environment. This hatred of Jews by their compatriots was exposed in its ugliest form in the hands of Hitler and Himmler, Goebbels and Goering. In the blood—chilling carnage of hapless Jews in Ouswitzs the world shrinks in awe and horror till date. But the root cause of the genocide lay several centuries back when the original inhabitants of Palestine were driven out by the newly emerging forces of jingoist Islamists.

Next target was the Babylonian people and Iraq and the Parsees of Iran. The latter were mostly converted, and the rest were annihilated and others were expelled. The campaign smoothly forwarded to the East till the frontiers of western Indian. Islam swept across all the areas upto Afghanistan. The porous Western border of India has all along been the easy entry point for outsiders. But here because of the presence of a very old and rich civilization, though degraded and decadent they had to adopt a different strategy. A sustained and thousand years' war was declared. Mahmud of Ghazni attacked and destroyed the Somnath Temple seventeen times. Hundreds and thousands of Hindu Temples were either destroyed or converted. Kalapahar a Hindu immediately after conversion to Islam took an oath to destroy Hindu and Buddhist temples. We do not find any trace of temples erected during the Pal and Sen Rajas in Bengal.

State of other states

*In Kerala the situation is such that any murderer, rapist or criminal can go scot-free if he converts to Islam or Christianity. T Nazeer, accused in several cases, is trying to.convert several Hindu prisoners to Islam. Conversion has forced seven girls of Sasthankotta in Kollam district to attempt suicide. Seven Hindu girls studying from 7 to 11 in Chakkuvalli Government School were studying in the SC hostel nearby. On July 23, they didn't go to school. When search was mounted for them by afternoon, by the school and the hostel authorities, it was found they were admitted in a private Christian hospital, in serious

---

*   (Facts collected from reliable sources)

condition, after consuming poisonous seeds. Recently two lakh people were rendered homeless, one thousand houses burnt to ashes and properties worth several crores destroyed in Kokrajhar and Dhubri Districts of West Assam. It is reliably learnt that Bangladeshi Muslim migrants have been preparing for long to show the combating and striking strength. Kokrajhar has a population of around eight lakhs of which a little over one lakh are reported to belong to Muslim migrants.

The fear of being swamped by illegal Bangladeshi infiltrators, who have been getting easy ingress from other parts of Assam, also exists among the Bodo Hindus as with the other communities in Assam. The Muslim infiltrators pose a serious threat to the very existence of original Bodo inhabitants who are by and large Hindus, a few among them being Buddhists. Another point of notable distinction is that *Muslim migrants have been used in the similar way by foreign forces to turn Assam into a Lebensraum. These Muslim forces, who are well entrenched within northeast, have created their armed cadres to suit the insurgency prone region. And it is alleged that armed Muslim cadres are used to foment the clash. The foreign force is alleged to instigate the migrant Muslims on communal line. It is strange to note that when Assam is burning because of attack from Bangladeshi Muslim migrants, the national media has just reported casually. The government too appears very casual in their approach.* (P Chidambaram took recourse to sermons like *"There are people from a variety of communities living in Assam now. Ultimately people of all communities would have to learn to leave together in peace."* He did not talk about border fencing to check infiltration.

Where are the Human Rights activists? It seems they don't find any violation of human rights of Bodos by Muslim infiltrators and by the state sponsored Muslim terrorism.

The state government has calculably refrained from taking any action to contain the Muslim influx simply because it is their vote bank. Swapan Dasgupta writes,

"That the origins of the violence lie in the demographic upheaval Assam has been witnessing for the past 100 years is undeniable. Thanks to the waves of immigration from the region that is now Bangladesh the population of Assam increased from 3.29 million in 1901 to 14.6 million in 1971, a 343.7 per cent increase compared to the all—India increase of nearly 150 percent in the same period. Public intellectuals in Assam have stressed that the increase of the Muslim population has been disproportionate. In an unusual intervention last week, the election commissioner, H. S. Brahma suggested that the details of the 2011 census may reveal that 11 of the 27 districts of Assam now have a Muslim majority . . . The Bodo-speaking minority which account for only five per cent of the population, perceives a dual threat from the Assamese-speaking majority and a physical challenge from Bangladeshi Muslims who constitute the majority in Dhubri and whose presence is increasingly being felt in the Bodo heartland of Kokrajhar district The emergence of militant Bodo sub-nationalism in the 1991s was an attempt to cope with these twin challenges and led to the formation of the semi-autonomous Bodoland Territorial Council in 2003. However much of the political gains from militant identity politics have been offset by the growing assertiveness of the Muslim community. The rise of the All India United Democratic front led by Maulana Badruiddin Ajmal , The All Assam Minority Students' Union and The Asom Mia Parishad has triggered a frontal Bodo Muslim confrontation. Tensions have further risen

following the AIDUF demand that the BTC be abolished because Bodos no longer constitute a majority in large areas governed by it. In an astute move, Ajmal has taken care to develop links with major Muslim organizations throughout India to ensure that the concerns of his social base are easily translated in "national" Muslim concerns. The government concerned with Muslim support has little scope to manoeuvre. Ground reports suggest an ongoing process of ethnic cleansing. Bodos in Dhubri are moving to Kokrajhar and dispossessed Muslims of Koktrajhar are moving to Dhubri. Some may even find their way into West Bengal. India's liberal intelligentsia have been very vocal on the so-called communal questions, especially the minorities. Yet the suspects have been strangely quiet over this monumental upheaval that has shaken Assam . . . . In 2004 when the religious demography of the 2001 census showed some strange results for Assam, the intelligentsia buried its head in the sand and ensured that all meaningful discussions on the subject were guillotined. The same process is once again at work over recent events in Assam."

The Telegraph, Aug 3, 2012

Dr Jay Dubashi writes, *"London may soon have more mosques than churches . . . . bearded young men everywhere, collecting donations for their mosques, and the police doing nothing about it, though begging is a crime in most western countries. The Europeans do not even realize they are slowly being deculturised, which is the first towards extinction as a race. This is what secularism has done to Europe. It has destroyed, or is about to destroy a whole civilization, which was once a fortress of Christianity and where religion has now become a dirty word."* We do not take his words on the face value. But it has definitely pointed out a very interesting event which has gone unnoticed

Religiosity has both good and evil effects. On one hand the general public can be regulated in a disciplined life and on the other it leads to narrowness and convert people into brainless bigots. But here secularism that is basically a belief that the state, morals, education etc. should be independent of religion has made people professing that opinion apathetic or abhorrent to any utterances of religious topics. Taking advantage of that indifferent attitude on behalf of the European Christians, Muslims especially and other religionists also are creeping in their countries from Asia and Africa, convert Christians and also Jews. So instead of being secular Europe is slowly but surely heading towards Muslim orthodoxy. Among the several reasons for the collapse of European civilization, of course economic decline and financial crisis cannot be overlooked, is the indifference to religion. This is a paradox.

Bengal Again: And the same paradox is being staged in Bengal since the last five hundred years. In the liberal atmosphere created by the resurgence heralded by the Bhakti Movement of Chaitanya Hindus of Bengal were so liberal they were mixing socially and culturally with the Muslims. But the rigid caste-based society and there being no system for baptizing in Hindu fold or reconversion once converted to Christianity or Islam a Hindu could not be taken back to their religion. On the other hand Islam does not permit any proselytisation. If anybody dares do so he would be killed. So it was only a one way traffic. The liberal Hindu, gentle Hindu, mild Hindu could be easily converted to a Muslim by matrimonial alliance, settled or negotiated or voluntary adoption of Islam. There are certain techniques for increasing the Muslim population:

Posing as liberal benefactors and participating in Hindu festivals and ensuring the trust of Hindus. And finally sneakishly convert them.

Approaching as genuine lovers marry a Hindu girl or boy and invariably convert him to Islam.

Creating barriers in their normal functioning of life.

Compel them to adopt Islamic way of life especially the conservative and retrograde ones

Intimidation and coercion

Beginning from 16th August, 1946 the Direct Action day declared by M A Jinnah the onslaught on Hindus have been continually going on.

Larry Collins informs—*"At dawn on 16 August, howling in a quasi-religious fervor, Moslem mobs had come bursting from their slums, waving clubs, iron bars, shovels, any instrument capable of smashing in a human skull . . . . They savagely beat to a sodden pulp any Hindu on their path and stuffed their remains in the city's open gutters. The terrified police simply disappeared.* Soon tall pillars of black smoke stretched up from a score of spots in the city, Hindu bazaars in full blaze . . . . Like water-soaked logs, scores of bloated corpses bobbed down the Hooghly River towards the sea. Others savagely mutilated, littered the city's streets. Everywhere the weak and helpless suffered most . . . . By the time the slaughter was over Calcutta belonged to the vultures. In filthy grey packs they scudded across the sky, tumbling down to gorge themselves on the bodies of the city's 6000 dead. The Great Calcutta killings*

---

*   The police did not get terrified. They were held back by Suhrawardy, the then prime minister of undivided Bengal. The police was asked to remain in barracks while the vultures will pounce on their prey.

as they became known triggered violence in Noakhali . . . . They changed the course of India's history. The threat the Moslems had been uttering for years, their warnings of a cataclysm which would overtake India if they were denied own state, took on a terrifying reality.

Jinnah vowed *"We shall have India divided, or we shall have India destroyed."*

Three crores of Hindu Bengalis had to leave the Eastern part of Bengal that is known as Bangladesh at present since 1971 (Here an explanation is required. Entire Bengal was known as Bangladesh to the ethnic Bengalis before partition. The Eastern part of Bengal i.e. original Bangladesh after partition became a part of Pakistan and known as East Pakistan. The western part of Bengal remained with India with the name of West Bengal. In 1971 the East Pakistan seceded from Pakistan and declared independence with the name of Bangladesh; .i.e. only a half of original Bangladesh is known as Bangladesh which is known to the international community as such. But strangely enough the majority Hindu Bengalis did not find any incongruity in this misnomer). Unlike in Punjab where there was complete exchange of population Muslims were not driven out of West Bengal in retaliation. Hindus and Sikhs left West Pakistan (presently Pakistan) While Muslims of Punjab left for West Pakistan. But no Bengali Muslim was forced to leave West Bengal. Only ten lakhs of Muslims shifted to East Pakistan only to fill the vacant space created by the fleeing Hindus to capture their properties, jobs and professions.

Since that fateful 16th August 1946 the **blackest** day in the annals of human history when war was declared on innocent unarmed ill-fated Hindu Bengalis the atrocities on them have been continuing unabated. (May be there were some non-Bengali Hindus, but their number was insignificant as most of them had an alternative place to take refuge in). Gandhi, the apostle of peace and prophet of Ahimsa and the pedantic and left oriented socialist Nehru were mute spectators to the outrage and trample of humanity. This was more agonizing as the propaganda for this mass carnage was going on for two decades. And the stalwarts of our freedom struggle (?) were in deep slumber knowing full well the outcome of this fanaticism.

In Noakhali district of erstwhile East Bengal (present Bangladesh) not a single Hindu woman was spared from being raped. The brutal killings and tortures of men folk made those who could avoid death run away. The women were spared and subjected to rape by adult Muslims with the definite target of producing an offspring who will naturally be a Muslim. Those women who declined to succumb to their monstrous design and loathsome scheme met with a cruel and painful end. Thus the purpose of increasing the Muslim population was served in this ghastly murder. Gandhi's presence could produce little change in the overall cleansing of Hindus.

The onslaught on Hindus in Bangladesh i.e. East Pakistan i.e. East Bengal came in waves in several stages with the continuous harangues in the intervening period. In 1948 all the trains, buses and aeroplanes were halted and the border sealed. The state sponsored massacre of Hindus went on for seven days on end. At long last Indian government secretly intervened and ultimately Nehru-Liakat Ali pact was signed. The residual Hindus living in East Pakistan have all along been made the sacrificial lamb on any pretext. The next wave came in 1964 when again Hindu Bengalis in East Pakistan were forced to leave. I was a school student then in the lower class. One early morning in winter I was awaken from sleep due to some noise. What I and others came to learn that a large number of Hindu families have arrived from the then East Pakistan. On enquiry we came to know that lakhs of such refugees have been squatting on the roads, stations and open fields. Z A Bhutto declared a thousand years war with Hindustan in 1965 not for nothing.

But the more heinous crime took place and has been on after 1971. The liberal Hindu Bengalis and common masses fondly started believing that with the creation of Bangladesh secular democracy will emerge in a Muslim dominated country. But to their disappointment their only hope Mujibur Rahman was brutally killed by Islamist forces along with his family members. The lone survivors were his daughter and her family members.

A new technique was devised to bolster their campaign. They started sending Muslim Bangladeshis in thousands and the then left government in order to inflate their vote-banks joined the fray. Ration cards were issued to these illegal Muslim immigrants through Panchayats and whoever are there in Panchayat the words of Muslims are final. These ration cards are accepted to be indisputable proof of their Indian citizenship. Nobody dared challenge the entry of these Muslims. As an Executive officer of Panchayat Samity I had the privilege to oversee and access to these gross violation of all kinds of legal, moral or humanitarian propriety. (Hindus crossing over to India have a pliable reason to do so as they are forced to in order to protect their life physically and culturally; but the same theory cannot be advanced for the Bangladeshi Muslims spilling over to India).

These same groups are now a formidable force to reckon with in the border districts of Bengal. In keeping with the tradition of Islamist expansion policy the day is not far when they will raise a war cry to secede from India and form a separate nation. Amongst these illegal immigrants a sizeable number have joined fundamentalist groups like SIMI, LeT, JeM, IM, and even Al Qaida.

In 1979 one retired Hindu school teacher in a village in Chanchal Block of Malda district of West Bengal made a complaint to the President of India. The complaint letter came to the District Magistrate through the Governor of West Bengal. The District Magistrate was instructed to conduct a thorough investigation into the matter and take remedial measures. The District Magistrate in turn ordered a magisterial enquiry into it and I was assigned the job. I proceeded with the Officer-in-Charge of Chanchal Police station to the village. The complaint was of a grievous nature.

He used to live in a village with a high density of Muslim population. His neighbor, another retired school teacher, with the aid of his three sons and other Muslim neighbors had been torturing the Hindu teacher so much so that it became impossible for him to live there. What did we see? The school teacher's tiled rooms were bearing ample proof of regular stone pelting. His two daughters still unmarried could not come out of home as they were continually harassed and their womens' modesty was outraged. They could not even draw water from the only tube well in the village. We collected evidences and witnesses.

The complainant was trembling as the fact of his making complaint was exposed and he was afraid of the repercussions. Very few people were bold enough to witness in black and white. I submitted a detailed report to the District Magistrate and recommended for initiating action under CrPC and IPC. But the report was shelved under pressure from political parties both in the government and in the opposition for fear of Muslim backlash and break down of law and order.

At the time of partition some Bihari and UP Muslim settled in East Pakistan. There number was around two lakhs. These people were not accepted by the local Bengali Muslims. On the eve of 15 th August, 1979 they declared that they would enter India through border areas on 14th August (it is the Independence Day of Pakistan) and settle in India in their original homes. The Indian government raised an alarm and asked the district authorities to ensure that not a single such person enters through border in their respective districts. At Border Security Force check post at Habibpur, Malda one team was formed to carry out the operation. We patrolled the entire Border throughout the day with the Commandant of Border security Force and The Inspector General of Central Reserve Police with me in overall charge under the instruction of the District Magistrate. Ultimately no such incident occurred and the administration felt relieved that no major law and order problem took place.

But that was actually the beginning of a process. The process of exporting foreign Muslims through the porous Eastern and North Eastern borders started. Now they have inundated West Bengal and other states as well. There is a demographic change and West Bengal has become a safe corridor for movement of terrorists and a haven for extreme fundamental groups. Hindus in Pakistan are in minority and they face atrocities at the hands Islamic majority there. Hindu girls are the subject of harassment. They are abducted, raped and forcefully converted and Hindu businesses are shut off and consequently Hindus are fleeing in lakhs. From Bangladesh more than three lakhs Muslims entered India with visa never to return. In addition to this there are constant illegal entries from Bangladesh.

On 14th August the mob in Mumbai hoisted Pakistani flags shouting anti-India slogans. They mutilated the martyrs' column wielding iron bars and throwing stones. They kicked at the memorial, smashed it with a lathi and then damaged the rifle and helmet inside the fibre glass casing. They took permission of 1000 people's rally but all Muslim groups gathered 50,000 strong mobs in Azad Ground. Mumbai assembly building, CST, Mumbai Corporation building, Naval Office and dock, prominent hotels are very near the spot. 50,000 thousand protesting Muslims delivered violent speeches, circulated abusive posters and then attacked Mumbai Police, media, buses and private vehicles.

It was nothing but a kind of treason and an attack from, in all appearances, another country. It was a war against the Indian nation from its proclaimed citizens. The same way Kashmiri separatists attack Indian's Army and police, the same way Kasab and his Jehadi friends from Pak attacked and killed Mumbai police officers and others on 26.11.2008. This was in support of Bangladeshi infiltrator Muslims in Assam and the Rohingigya Muslims in Myanmar, infamous for their jehadi terror links from Afghanistan to Thailand. They molested 3 woman constables and also abducted a policeman. Foe 1400 years a process of Islamisation is going on all over the globe. The method is: First attack and occupy a part of a land and rule that land. Persecute and chase away original or more correctly the non-Muslims inhabitants until and unless they adopt Islam. Those who oppose perpetrate torture as crude as possible, rape their women, carry out ethnic cleansing. Then slowly infiltrate the neighboring territories ;increase the population; claim to be under-privileged ; demand concessions for 10-12 per cent vote banks grab facilities over the majority and with ,ironically , majority's money;

create pressure on government , society and globally; then again start attacking non-Muslims wherever they form a sizeable section 1-15 per cent. Example: Kashmir, Assam Andhra, Kerala, Bengal in India. Old Paris in France South Myanmar and Thailand.

(Collected from reliable sources,—famous international journals, police records, number one English dailies in India and abroad)

———◈———

It matters little which religious practices one professes. Religion has no other significance than controlling the baser instincts of men and guiding them to lead a cool, composed and civilized life. As man is basically very restive and directionless he needs a well—defined way of life or else the people if allowed to behave wantonly there will be utter chaos and ultimately this society and hard-built civilization will crumble. As a species human race will be extinct. It was argued by someone that if we could read the mind of another man the world would have been destroyed within a few hours. The same principle does not apply to other species. There is no such animal that is so much scheming, calculating and self-seeking other than human being irrespective of any religion, language, caste, creed or sex. Because of our superior intelligence and its natural corollary the greed for power money etc. make every man the potential enemy of another man and which is why so many prophets and philosophers have indicated the solutions according to their realization to halt this proclivity. For someone who can assimilate the virtues preached by religion without falling prey to the vices often associated with orthodox religionists religion has little value.

But we cannot eradicate religion in any way. Sorry, Respected Marx, your 'opium' cannot be unseated from its prized position. It shall be there. If one religious community clamors more for the upholding and observance of their religion and threaten to pull out swords if slightest criticism comes from any quarter it appears all others including the mightiest political power in the world surrender to the threat. And since religion will not die and will be there to stay so every religionist should be treated on par and given the same opportunity to flourish, preach, propagate and percolate.

Course of events could have taken a different turn had the brighter sides of Islamic or, say, Arabic civilization built by the Arab alchemists and astronomers been given eminence; but here also the march was stopped by the orthodox forces. Islam does not divide society on caste which is a permanent stigma on Hindus. Islam teaches that this life is robust, colorful and to be enjoyed with all its diversity. But the spirit imbued with the resurgence in Islam was lost by recall of the petrified diktats of fundamentalists. Always the less intelligent majority are guided by a leader or group of leaders of their choice and of their status and the latter are on the pry to dethrone the rational and scientific intelligentsia. In the latter chapters this topic will be covered.

CHAPTER 9

Science and Religion

Despite vehement opposition from age old ignorance emboldened by religion science has grown. Man's insatiable desire to know the unknown and discover the cause behind each natural phenomenon made this possible. The often repeated claim by the theists and religionists that scientists believe in God has to be scrutinized.

In the ancient period the Greeks had the habit of observing the movements of celestial bodies and acquired some primitive knowledge about the position, size etc of the stars and the sun. During the dark middle Ages it was the Arabs who kept the torch alive. They cultivated science for philosophical inquisitiveness and for fulfillment of practical need of navigation. Arabs were adept in accurate observation and mathematical calculations. Interestingly they were not prejudiced about the position of the earth or sun and stars. They had no contagion of Catholic belief of geo-centric universe until sixteenth century. The Arabs had carried their bounties of knowledge to Europe. The Arabian scientists had imparted their discoveries to Europe. The reaction in Europe was a mixed one.

Roger Bacon the forerunner of modern science told; *Take nothing on trust.* That is why he proposed that no scientific formula can be accepted without experimental proof. Till this day any postulate of science is validated only if it is established by experiment. He studied mathematics, astronomy, optics, alchemy and languages. The gunpowder was manufactured from charcoal, saltpetre and sulphur. The Greek Marcus Graecus invented it. But with decline of Greek civilization the technique was missing. It was recovered by the great Arabian scientists whom Roger Bacon regarded in high esteem. He was the first European who reinvented gunpowder in improved version using the old method.

The primitive man used the bow to throw an arrow without knowing that in the process he has discovered an important principle of physics. The potential energy was converted into kinetic energy. In the same way the kinetic energy of wind was utilized to make the boat or ship to run against the flowing river without necessitating strenuous physical labour of man. The gunpowder proved that a small quantity of matter may contain huge amount of energy which is latent in a substance. The chemical energy is converted into mechanical energy. He made significant studies on the nature of light and on the rainbow and carried some experiments on lenses and mirrors. He suggested that a balloon of thin copper sheet filled with 'liquid fire' can be made to fly in air as many light objects do in water. He even tried to fly with flapping wings. Bacon described spectacles which also came into use soon, clarified the principle of reflection, refraction and spherical aberration and mechanically propelled ships and carriages.

He used a pin whole camera to observe solar eclipse. He wrote:

Because these things are beyond your comprehension, you call them the work of the devil; your theology and theologians and canonists abhor them as the products of magic, regarding them as unworthy of a Christian. He criticized severely the theologians and the Christian scholars of his time. He was despised for his dependence on alchemy and astrology. But even this rebel had to hide behind the Christianity. Sometime between 1277 and 1279 he was imprisoned. But how long he was in prison is not known.

Copernicus: He was dissatisfied with the Earth-centred ideas of the universe and spent 33 years quietly meditating on this. He developed the ideas already prevalent and declared that the Earth and other planets revolved at a point in space round the sun. Man was, according to Christian belief, created after the image of the God. Now the Earth lost its prime position of being the centre of the universe. With Copernicus' discovery it was only a minor planet revolving along with other planets round the sun. Hence man no longer enjoyed its supreme position that in turn meant the indignity of God.

A great opportunity was laid bare. Very soon it was revealed that the sun was also not at the centre of the universe. There are numerous stars in the universe with their family of planets and satellites of which the sun was only a medium sized one. The major work of Copernicus, *On the Revolution of the Celestial Orbs was* published in 1540.The geo-centric vanity also took away the ego-centric pride. So far man was concerned only with his emancipation of soul. Now scientists were endowed with a great power for discovering newer things that provided him with the capacity for acquiring real knowledge.

This knowledge unfolded before him hundreds of opportunities since it opened the floodgate of discoveries in science and helped man to harness latent energy in matter to utilize it for practical purposes. Four hundred years ago it was unthinkable to question the authority of the church. It was considered a grave crime. Copernicus was reluctant to publicize his revolutionary ideas afraid of cruel punishment. Here comes the intervention of the Church. Schomberg, patron of Copernicus and a Church priest appended that it was only a hypothesis. Copernicus wrote in preface

That I too began to meditate on the motion of the earth, and though, it appeared an absurd opinion, yet since I knew that in previous times others had been allowed the privilege of feigning what circles they chose in order to explain the phenomena, I conceived that I might take the liberty of trying whether, on the supposition of the earth's motion, it was possible to find better explanations than the ancient ones about the revolutions of the celestial orbs.

It is clear to note that the oppressive social milieu and religious fanaticism forced Copernicus to disown his own great revolutionary discovery. He had to publicly admit that he his view was **absurd** and it was considered as **privilege to feign** to think otherwise than the church believed. He also asked for that privilege not as a natural right for freedom of expression but as it has been allowed to others in previous times. Copernicus died only a few days after his work was published. The Church would have served a dreadful penalty to the old scientist had not death intervened. Human history escaped another act of barbary. We can quote the following from his writings

If there be vain babblers who, knowing nothing of mathematics, yet presume to pronounce judgment, through an intentional distortion of any passage of the scriptures, and, who, blame and attack my undertaking, I heed them not, and look upon their judgments as rash and incomprehensible.

Bruno: In 1584 Giordano Bruno not only reaffirmed the reality of the helio-centric theory but suggested that the universe is infinite, constituted of innumerable worlds similar to solar system. He also preached that Bible should be followed for its moral teaching but not for its astronomical implications. He formulated that religion is considered as a means to instruct and govern ignorant people while philosophy is a discipline for the elect who are able to behave themselves and govern others. It may be pointed out here that science was then considered as philosophy and that is why it was called natural philosophy.

His *Expulsion of the Triumphant Beast* is a satire on contemporary superstitions and vices. **It was a scathing attack on the Calvinistic principle of salvation by faith alone**. In his next book *Cabal of the Horse Pegasus*, published in 1585 he negated the absolute individuality of human soul isolated from the universal soul. In *The Heroic Frenzies he* treats of union between the Infinite One and the human soul and exhorts man to the conquest of virtue and truth. Bruno entered into a polemic with a protégé' of the Catholic party, the mathematician Fabrizio Mordente and ridiculed him. He even attacked Aristotle. In 1588 he published "160 Articles" he enunciated his view of religion. He preached peaceful coexistence of all religions based upon mutual understanding and the freedom of reciprocal discussion. Bruno was arrested in 1593 and tried. The trial ran for seven years **he displayed no interest in theology but stressed the philosophical character of his ideas**.

He was not a scientist in a technical sense but he represented the spirit of science that came as a terror for religion. The age was suspicious of anything new and enlightening. Bruno wrote that he found "scepticism under the polish of hypocrisy" He was "not against belief but against pretended belief." His remarks which irked the Orthodox Church are:

Space was infinite and it was filled with self-luminous bodies.

The infinity of forms under which matter appears, it does not receive from another and something external, but produces them itself and engenders them from its bosom.

Matter is not without form . . . it contains them . . . within itself; it is . . . all nature and the mother

---

Without the materialistic view of the universe no science would be possible.

The infinity of forms under which matter appears, <u>it does not receive from another and something external, but produces them itself and engenders them from its bosom.</u>

<u>Matter is not without form . . . it contains them . . . within itself, it is . . . all nature and the mother of all living things</u> This was an atomic basis of matter and being.

1591 he went to Frankfurt am Main where he was not permitted to stay. He earned the reputation of being a "universal man "who did not possess a trace of religion In 1592 Mocenigo initially showed interest in Bruno's philosophical investigation without its theological implications. But finally he was disappointed and denounced him to the Venetian Inquisition in May 1592 for his heretical theories. Bruno was arrested and tried. He defended himself by admitting minor theological errors, emphasizing however the philosophical rather than the theological character of his basic tenets. The Venetian Inquisition was favorable to him but the Roman Inquisition demanded his extradition and on Jan27 1593, Bruno was imprisoned. During the trial that ran for seven years he disclaimed any interest in theological matters and reaffirmed his conviction in philosophical character of his speculation. Dissatisfied Church demanded an unconditional withdrawal of his views. He made a frantic effort by holding that his views were not incompatible with the Christian conception of God and religion. But The Inquisitors pressured him to issue a formal disclaimer. Bruno finally declared that he has nothing to retract and he did not know what he was expected to retract. Pope Clement VIII ordered that he be sentenced as an impenitent and pertinacious heretic. On February 8, 1600 when the death sentence was announced he addressed his judges, "Perhaps your fear in passing judgment on me is greater than mine in receiving it." He was brought to the Campo de' Fiori, his tongue in a gag, and burned alive. As historian Draper has stated," Never can moral interests, however pure, stand against intellect-enforced truth. On this ill-omened question, the Church ventured her battle, and lost it." He was served the severest of punishment that can be as exemplary as it is required to cower any scientist in future to question the authority of God or Church.

[We have not gone into the details of this grand and majestic life, rather we kept ourselves confined to the salient features of his pronouncements that brought him into clash with the orthodox Christian church.]

Medical Science: The Earth which was imagined to be stationary now came to be known as moving. The slaughter of Bruno did two things. It established that the abiotic world or the inanimate world can be explained with the help of mathematics and nothing comes out of nothing. The matter is not to be created. It changes form only. With the help of mathematics and the new found Astronomy the mysteries of the inanimate world can be studied. Now human mind also began to penetrate into the domain of mysterious human body and trying to ascertain the laws of human body.

*Hippocrates and Aristotle regarded the heart as the source of blood, blood vessels and an innate heart which caused pulse and heart beat. Galen (AD. 129-199) proved by vivisection that the left ventricle contained blood, but thought that it passed to the right ventricle through invisible septum. The heart beat propelled blood through the arteries from the left ventricle, while the right allowed waste fumes to leave. The fallacy was first exposed by the Arab physician Ibn an—Nafis(AD 1205.-1288) who showed that blood travelled from the right ventricle to the left via the lings , but his ideas were not assimilated and later were forgotten. Leonardo da Vinci's (1452-1519) anatomical drawings of the heart valves could have helped to dispel the errors of the ancients. But they were privately owned and not widely studied. Andreas Vesalius (1514-1564) noted that the septum between ventricles appeared impenetrable, but offered no explanation how blood travelled from the left side of the heart to the right. **Michael Servetus (1511-1553) stated that blood coursed from the right of the heart to the left via the lungs, but he was burned for heresy**. Later in the same century, the tide began to turn.Andreas Cesalpino (1519-1603) who coined the word circulation stressed the return of venous blood through the veins, and*

Harvey' teacher at Padua , Fabricious ab Aquapendente (1533-1619) taught that Valves in the veins prevented venous blood from rushing downwards. Stephen Hales (1677-1761) measured blood pressure by inserting tubes into the arteries and veins of animals and observed the height to which the blood rose and its variations with the heartbeat(The World of Science.)

Aristotle was the supreme intellectual in Europe for about two thousand years at least in material issues. He dominated the world of knowledge. Similarly was the position in respect of disease, medical treatment and physiology that was dictated by whatever Galen (129-199 AD) another Greek taught. It was considered heretical to oppose Galen's theory until the sixteenth century. The University of Salermo was the main Medical College. It is true that Galen had pronounced some fundamental principles of medical science. But like original religious scripts which were very much precise and substantive were in course of centuries mixed up with superstitious belief because of so many factors e.g., lack of control, and tendency of people of lesser intellect to show off their caliber, Galen' s knowledge were spoilt with many superstitious belief due to lack of opportunity of proper experiment. We must remember that the progress of medical science was cotemporaneous with that of physics and chemistry. Hence with these two branches of science biological science also developed as instruments for verifying various factors concerning human body were now available like temperature, blood pressure, pulse rate cells count, detection of bacteria, germs and virus so on and so forth.

Paracelsus was the first to question the teachings of Galen and with existing knowledge of Chemistry he utilized for service of medicine. His audacity was to teach that nature was the greatest healer and that the function of medical science was to help the normal functioning processes inherent in the structure of human body. He regarded the human body as independent of any soul as proclaimed by religions and theologists. He performed dissection of human body in presence of students and showed them that what Galen observed was wrong and asked his students that if his findings were at variance with that of Galen, the latter should be ignored. That added fuel to the fire. This time the resistance came from classical medical learning and not from any church. This challenged the traditional ignorance—based pretended knowledge. It was believed that every disease and cure of the human body was predetermined as a part of cosmic law that there is some relation between macrocosm and microcosm. The findings of Vesalius established that diseases are not predetermined. By dissecting human bodies it was proved that the weight of the living body and dead body was the same and no soul came out of the body. And they were not caused by the curse of God. Diseases can be classified and its cure is possible by the application of proper dose of medicine and observance of some fundamental rules of hygiene

Galileo: Galileo Galilei within nine years of execution of Bruno discovered telescope (though Lippershy, a Dutch scientist independently discovered telescope in 1608). He placed two lenses at the ends of a cylindrical tube. Bruno had already declared that the sun was at the centre of the solar system and the blue dome consisted of many fixed stars of which the sun was one. The stars were not there to illuminate the night as designed by God. Galileo identified forty fixed stars. He informed that they were not fixed but so far away that only a small portion of its light reached the earth. And their position seemed to be fixed as they were very far so that their displacement from the earth were not perceptible. Moreover if they were only nearer by ten percent the light would have been so much blazing that animals including man would have been blinded . So the story of stars of being created by God for pleasure of man instead were there as a part of macrocosm. In the winter of 1609-10 Galileo saw the mountains of the Moon and the four bright satellites of the planet Jupiter and his observations confirmed his already strong belief that the Sun was the centre of the System. The satellites of Jupiter were particularly important; they proved there were more than one centre of movement in the planetary system.

According to the old Ptolemaic theory Venus could not possibly show a full sequence of phases from new to full. But Galileo observed that it Venus also showed its phases. Galileo published these observations in "Sidereus Nuncius" (The Starry Messenger). It received wide acclaim but some churchmen were skeptical. He was welcome by the Pope. As he moved from Venice to Florence he found the atmosphere not that congenial. In Rome one Cardinal, Bellarmine wrote that the theory of a moving earth " . . . injures our holy faith and make the sacred scriptures false." In 1616 Pope himself intervened and through Bellarmine **ordered Galileo to cease teaching the heresy of a moving earth**. But Galileo's old friend Barberini became Pope Urban III. In 1630 Galileo published book in the form of a dialogue between characters, Salviati who supports the Copernican theory, Simplicio, who opposes it and Sagredo, who is more or less neutral. Simplicio is made to look naïve and even ridiculous so that in effect the "Dialogue" is an open propaganda for the theory of Central Sun and a moving Earth. Galileo sent the manuscript to Rome where it was ultimately published in 1632 with the consent of the

Holy Office. But Barberini, the good old friend of Galileo, now Pope, was a totally different person. He read the book and felt that Galileo's Simplicio was a caricature of himself and he was ridiculed. The ageing scientist was called to Rome, arrested on a clearly fabricated charge of heresy, tried and condemned. **In June 1633 Galileo was forced to recant-to "curse, abjure and detest" the false theory that the Earth moved around the Sun**. He was horrified by the painful death of Bruno and had to recant. But it was ensured that the old man does not go unpunished so that others get a message and dare not challenge the authority of the church and the established notions. The conflict between reason and foolishness, ignorance and knowledge, liberal humanity and dogmatic belief again came to the fore. He was not allowed to see his friends and relatives in prison. Finally he was allowed; by that time he was blind; towards the end of his life he could not even hear.

Descartes:

A series of discoveries shook the world of science between Galileo's laws of motion and Newton's laws of motion and universal law of Gravitation. Keplar deducted the laws of planetary motion. Lavoisier discovered oxygen, Boyle discovered law of gas pressure, and Torricelli measured atmospheric pressure and invented Barometer. Drebble manufactured thermometer, Gueric invented the air-pump. Roemer measured the velocity of light. Harvey discovered the circulation of blood.

Descartes established the Coordinate Geometry. But apart from his scientific discoveries he is remembered as a leading philosopher of the seventeenth century. He explained the theoretical process of scientific development. He held that for finding the truth we must doubt everything except the reality of his own doubt. Whatever I apprehend very clearly and distinctly is true. Nothing can result from nothing. Descartes made a concession to the idea of God because God was clear to him. His argument was: The physical world is real, simply because it is there. I exist because I think about it. If it did not exist, there could not be any thought about it. This argument obviously excludes the authority of God. The reality of the physical world thus guarantees not necessarily the existence of God, but only his morality. This is a matter of playfulness. **He tried to avoid clash with the clergy; the clever man tried to pull their legs**. He put forward the theory that the world developed from small particles with the observation that of course God has created the world at one time, but it was very interesting to see how the world might have developed by itself.

Johannes Kepler belongs to both the past and
present. Though a brilliant mathematician he was a mystic. And some his theories seem to be very strange. Kepler established three laws of planetary motion with the help of mathematics.

1. The planets move round the earth in an elliptical path with the sun at one focus. The sum of the distances from two foci is constant
2. The radius covers equal areas in equal times.
3. The square of the orbital period is proportional to the cube of the mean distance.

He believed five regular bodies existed between the sun and the five planets discovered till then. He was not subject to torture as Galileo had been but he never got a good job.

He practiced astrology but it is not clear whether he did it out of curiosity or he had much faith in it. To earn livelihood he started writing pamphlet on astrology. His three children suffering from pox died for want of money and medical treatment even by the standards of that time. His old mother was charged with witchcraft and condemned for no other offence than giving birth to a child who offended Christian religion. His old mother was imprisoned, he died of brain fever.

What is most remarkable here is to note that religion was only an excuse. It happens in every age that whoever attempts to break the existing ideas and replace old ideas with new ones are reprimanded by the people who are vastly read in the already established ideas but incapable to go deeper and unfold so far undiscovered truths. The individual and collective ego come in the way of accepting any one who is endowed with greater talent or genius to unravel anything outside of their knowledge. This '**learned ignorance**' is also a truth of humanity. The truth is that man is hostile to acknowledge the truth in the absolute sense. First we have to identify and agree to that realization. Some people who are over confident not to accept this most apparent fallacy and argue that man is not so foolish in the sense it is stated here befool themselves. We are in all ages and in every society not prepared to recognize a new idea that is not in consistence with the information already accepted by the majority and in currency over a long period. Whatever our forefathers have told are accepted doggedly. It is considered an aberration to speak in a different tone. Truth is not something that is established in an indefinite past and by some unknown but very powerful persons almost akin to God. It is in the continuous process of unfolding. Who knows where we will stop in that process but it is more enlivening to keep the process on. The doomsday shall come only when we get answer to all our queries and stop wondering.

Newton

But why do people seek asylum under the umbrella of religion? Because religion is the most powerful weapon and the effective intoxicant to rally the masses against anyone who throw challenge to the establishments. The power of religion is misused by the erudite but non-creative persons in league with the religious heads that are eager to keep hold on the masses in the easiest way. Galileo's Laws of motion set in motion the idea of this moving world. Bruno asserted that motion once generated does not cease. If it be so it had no beginning. Descartes reduced all forms of motion to the motion of fine corpuscles that comprised all the matter. He made no distinction between the organic world and inorganic world. "Give me matter and motion and I will construct the universe". Thus his theory is a rejection of all mystical explanation of nature. It threw challenge to all religious beliefs. On the basis of empirical evidence and on the strength of his predecessors Newton postulated his Laws of Motion and theory of universal Gravitation. On the assumption of truthfulness of his laws he deduced formulae which could be put to tests and found correct. He postulated a God which set the universal mechanism in motion. But once the motion was given to the universe it went on completely independent of Nature.

The frowning eye of Church and scholars could not be ignored. Hence the scientists had to give a formal recognition to the faith that the world started from an impetus given by God and the rest were governed by the laws enunciated by them. The faith in God had to be admitted because the scientists are not martyrs or fanatics to lay down lives for a cause good or bad. They simply want to carry on their investigation of truth of the material world without getting into clash with the powerful lobby in any way.

ABOUT THE AUTHOR

A National Scholar he graduated from Presidency College Kolkata and joined Civil Services as Deputy Magistrate & Deputy Collector. A student of Physics but a keen observer of historical events with its attendant social contours. He has a point to establish regarding the course of historical events. After all history is 'his' story that is the story of man.